CAVITATION

Cavitation is often present in fluid flows and fluid machinery, ranging from blood flows to turbines and ships' propellers. Essentially the low pressure boiling of the liquid, it depends for its formation on the presence of nuclei, on the temperature, pressure and speed of the fluid flow. Its effect on machines may be loss of efficiency or erosion damage to the blades, whilst in blood flows it may damage the arteries or the heart, as well as causing the "bends". It can also produce weak light emission called "sonoluminescence".

Its effect on hydrofoils is to cause loss of lift, and a similar effect is caused on the performance of machines, pumps, turbines and propellers. Each type of machine is examined in turn and the cavitation performance characteristic investigated, together with the design recommendation and optimisation. Cavitation erosion causes trouble in many machines and static devices. Methods of overcoming it are mostly by the use of stronger materials and extensive comparable tests give a good guide in this respect.

Some new devices attempt to make use of cavitation, either in the fluid machinery field (supercavitating propellers and pumps), or for chemical mixing or reactions.

Other titles in this series

ME/1 TRIBOLOGY
E D Hondros

ME/2 ELECTRON BEAM WELDING
M J Fletcher

ME/3 VACUUM BRAZING
M J Fletcher

ME/4 STRESS CORROSION FAILURE
P Greenfield

ME/5 GLOW DISCHARGE MATERIAL PROCESSING
R A Dugdale

ME/6 DIAMOND GRINDING
J Burls

ME/7 ENGINEERING APPLICATIONS OF BERYLLIUM
P Greenfield

ME/8 WHISKERS
C C Evans

ME/9 CREEP OF METALS AT HIGH TEMPERATURES
P Greenfield

Other Mechanical Engineering titles available

TL/ME/1 HYDROSTATIC EXTRUSION
J M Alexander and B Lengyel

TL/ME/2 THE HEAT PIPE
D Chisholm

All published by Mills & Boon Limited

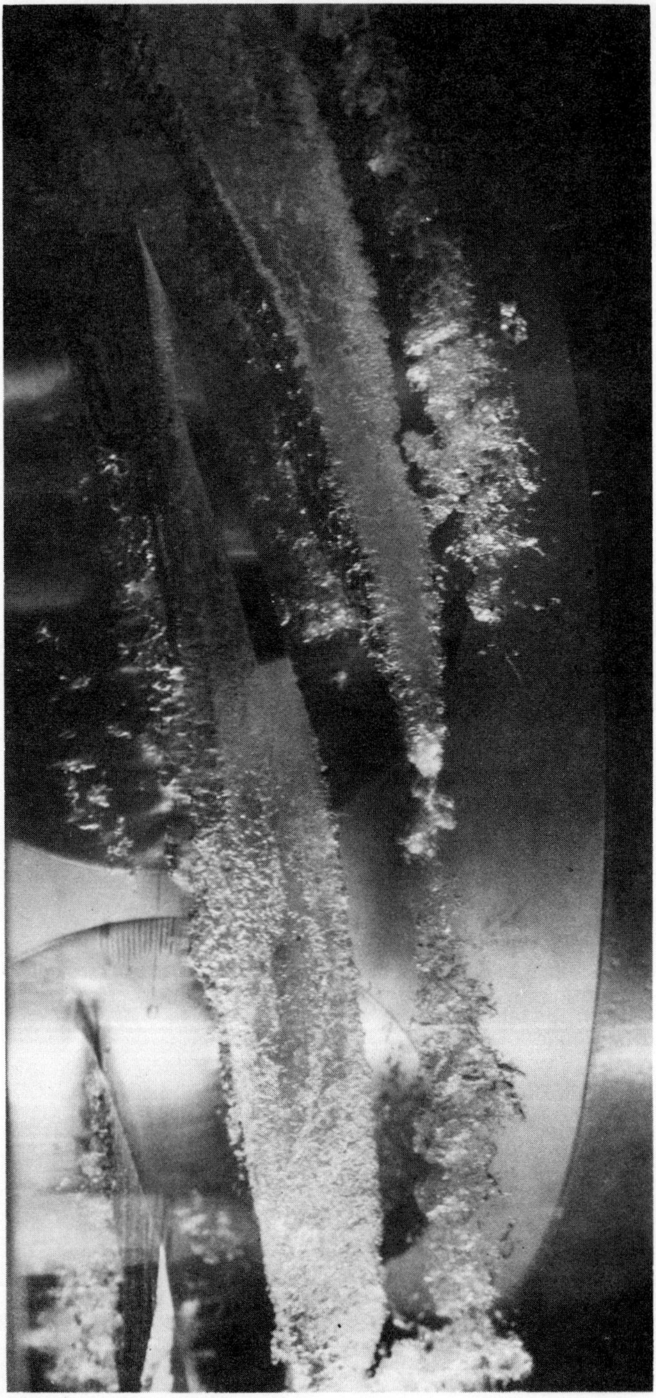

Frontispiece. Supercavitation.

M & B Monograph ME/10

General Editor: J Gordon Cook, PhD, FRIC

Cavitation

I S Pearsall, BSc (Eng),
PhD, MIMechE

National Engineering Laboratory
Fluid Mechanics Division

Mills & Boon Limited
London

First published in Great Britain 1972
by Mills & Boon Limited, 17–19 Foley Street,
London, W1A 1DR

ISBN 0.263.05000.9

Made and printed in Great Britain by
Butler & Tanner Ltd, Frome and London

CONTENTS

NOTATION

A	Swept area of propeller
B, B_1	Stepanoff thermodynamic parameters
C	Absolute velocity
C_L	Lift coefficient
C_P	Specific heat
D_h	Hub diameter
D_t	Tip diameter
D	Propeller diameter
h_v	$(= P_v/\rho)$
h_s	Suction lift
Δh	*NPSH*
Δh_f	Change of enthalpy
H	Total head (across pump or turbine)
J	Advance coefficient
J'	Mechanical equivalent of heat
k_1, k_2	Empirical coefficients
K_T	Thrust coefficient
K_Q	Torque coefficient
L	Latent heat
N	Rotational speed (rev/min)
n	Rotational speed (rev/s)
NPSE	Total energy at pump suction—vapour energy (= gNPSII)
NPSH	Total head at pump suction—vapour head
P	Static pressure
P_a	Atmospheric pressure
P_v	Vapour pressure
P_c	Critical pressure
Q	Flow
r_v	Ratio of volume of liquid and vapour
U	Peripheral velocity
V	Velocity
$\bar{V}$	Volume
v	Specific volume
Δt	Change in temperature
T	Temperature
W	Relative velocity
Y	Specific energy

Z	Level
ϕ	Flow coefficient (V_a/U)
ρ	Density
σ	Cavitation coefficient
σ_{Th}	Thoma coefficient
σ_b	Blade coefficient
η	Efficiency
β	Fluid angle
ω	Angular velocity (rad/s)
λ	Hub/tip ratio (D_h/D_t)
ζ_c, ζ_v	Empirical coefficient

Subscripts

0	In pipe before pump
1	In annulus before blades
2	Behind impeller
∞	Mean of 1 and 2
m	Meridional
a	Axial
u	Swirl
t	Turbine
s	Draft tube

Specific and Suction Specific Speeds

$$K_n = \frac{\omega\sqrt{Q}}{Y^{3/4}} = \frac{\omega\sqrt{Q}}{(gH)^{3/4}}$$ (rad/s, Q, H and g, in consistent units)

$$K_s = \frac{\omega\sqrt{Q}}{NPSE^{3/4}} = \frac{\omega\sqrt{Q}}{(gNPSH)^{3/4}}$$ (rad/s, Q, H and g, in consistent units)

$$N_s = \frac{\text{rev/min}\sqrt{\{Q\ (\text{gal/min})\}}}{H\ (\text{ft})^{3/4}}$$

$$s = \frac{\text{rev/min}\sqrt{\{Q\ (\text{gal/min})\}}}{NPSH\ (\text{ft})^{3/4}}$$

$$N_s' = \frac{\text{rev/min}\sqrt{\{Q\ (\text{m}^3/\text{s}\}}}{H\ (\text{m})^{3/4}}$$

$$s' = \frac{\text{rev/min}\sqrt{\{Q\ (\text{m}^3/\text{s})\}}}{NPSH\ (\text{m})^{3/4}}$$

$$N_{st} = \frac{\text{rev/min}\sqrt{(\text{metric } HP)}}{H\ (\text{m})^{3/4}}$$

1. Introduction

Cavitation had been postulated theoretically by Reynolds[56] long before it was first observed in the trials of the destroyer HMS *Daring* in 1893. The speeds reached by HMS *Daring* were well below those expected and the trouble was traced to poor propeller performance due to formation of vapour bubbles on the blades. A few years later the first turbine ship, the *Turbinia*, met similar difficulties. She was fitted with seven different propellers in turn but none enabled her to reach anything like the predicted speed. Eventually, it was found necessary to replace the single shaft by triple shafts with three propellers on each; thus equipped, the *Turbinia*, attained a speed of 32 knots. The development of these propellers was carried out in the first small water tunnel, built by Parsons Steam Turbine Co. Ltd.[57]

Cavitation is caused by the boiling of a liquid at normal temperatures and low pressure. It is accentuated by the presence of dissolved air which is released when pressures are reduced. It is the combination of air release and vaporisation caused by low pressures induced in a fluid that we call cavitation.

Since the time of the *Daring* and the *Turbinia* investigations, the chemical, physical and engineering aspects of cavitation have been the subject of intensive research. Investigations have been made of the process of bubble formation and collapse, the role of gases dissolved in the liquid, the effect of bubble formation and collapse on pumps, turbines, propellers, dams and similar structures. It has been found for instance that the formation of bubbles upsets the flow of fluid through a machine, causing loss of pressure, efficiency and thrust, as in the case of HMS *Daring*. Also, the collapse of the bubbles causes shock waves which erode the metal of the turbine impeller or propeller.

To investigate such phenomena, sophisticated test equipment has been built, ranging from large water tunnels of several thousand hp used for testing propellers or turbines, to small high-speed vibrating apparatus used in the investigation of erosion damage.

Most aspects of cavitation have been investigated at one time or another, but solutions of the major problems have not yet been achieved. Sufficient is known about cavitation, however, for engineers to be able to live with it.

2. Cavitation Inception

2.1 INCEPTION AND NUCLEI

In theory, a liquid will vaporise when the pressure is reduced to its vapour pressure. In practice, the pressure at which cavitation starts is greatly dependent on the liquid's physical state. If the liquid contains much dissolved air, then as the pressure is reduced the air comes out of solution and forms cavities in which the pressure will be greater than the vapour pressure of the liquid. Even if there are no visible air bubbles, the presence of submicroscopic gas bubbles may provide nuclei which cause cavitation at pressures above the vapour pressure. Each cavitation bubble grows from a nucleus to a finite size and collapses again, the entire cycle taking place perhaps within a few milliseconds. Bubbles may follow each other so rapidly that they appear to the eye to form a single continuous bubble.

In the absence of nuclei, the liquid may withstand negative pressures or tensions without undergoing cavitation. In theory, a liquid should be able to withstand tensions equivalent to thousands of atmospheres. It is estimated that water, for example, will withstand a tension ranging from 500 to 10,000 atmospheres.(1) In practice, even when water has been subjected to rigorous filtration and pre-pressurised to several hundred atmospheres, it has ruptured at tensions of 300 atmospheres. According to Plesset(3) water without nuclei theoretically will withstand tensions of 15,000 atmospheres, but the probability of this happening is low unless the bubbles are of molecular dimensions. However, when solid non-wetted nuclei of size 10^{-6} cm are present, it is likely that water will rupture at tensions of the order of tens of atmospheres. It follows from this that it should be possible to raise water under vacuum to a height greater than that corresponding to atmospheric pressure. This has actually been achieved by pre-pressurising the liquid.(2)

Even when local pressure conditions are known accurately, it is far from easy to predict when cavitation will occur, because it is necessary to estimate the distribution and size of the nuclei present. An approximate answer can be achieved by measuring the liquid's total gas content, as this can be related to the pressure at which cavitation occurs. Nevertheless, until research yields a better understanding of the role of these nuclei, prediction cannot be made exact.

If nuclei in the form of very small bubbles are to exist, some stabilising influence must be present. If the bubbles are very small, surface tension effects should make them collapse; if they are large enough to be visible they should rise to the liquid surface and escape. Various hypotheses have been suggested to explain this inconsistency; for instance, the nuclei may be trapped in minute cracks in the fluid's boundary surface. This is substantiated to some extent by the fact that cavitation usually starts at or near such boundaries. Nevertheless, cavitation can be induced far from a containing wall—for example, in the centre of a vortex or at the focus of an ultrasonic fluctuating pressure field. If solid particles are suspended in the fluid, then the "surface cracks" hypothesis could still hold, the particles replacing the walls as centres for the nuclei. This suggestion has considerable relevance, as Plesset(3) has shown. Another suggestion is that the bubbles are surrounded by an organic skin formed from impurities which act as a stabiliser. Existing experimental evidence does not support this hypothesis. In special cases, nucleation may be caused by charged particles, such as cosmic rays.

Cavitation gives rise to other physical phenomena. During the final collapse phase there is a weak emission of light from the bubble. The cause of this sonoluminescence is not known. It has been suggested that the flash is due to the recombination of free ions produced by thermal or mechanical dissociation of molecules on the bubble surface. On the other hand, Jarmen and co-workers(53) in 1960 put forward strong evidence to indicate that sonoluminescence is due to the incandescence of the gas in the bubble resulting from the high pressures and temperatures during collapse. The flash may last for a period of between one twentieth and one thousandth of a second. The intensity of the light is dependent on the gas present; there will be no light if there is no gas in the bubble. Sonoluminescence is only visible by magnification or in complete darkness, as the emission is very weak. Sparks have also been seen during the cavitation process, usually in oils, and it has been suggested these are due to static discharges.

When the cavities collapse, high pressures and temperatures are caused in the bubble. Surrounding fluid temperatures as high as 10,000 °K have been suggested. Wheeler(4) has concluded that temperature rises of 500 to 800 °C can occur in the material adjacent to the collapsing bubble. The collapse is completed in a period of milli or even microseconds. Harrison(54) showed that the resulting shockwaves radiated through the liquid adjacent to the bubble may have a pressure difference as high as 4000 atmospheres.

The life of a cavitation bubble thus involves two changes of phase separated by expansion and collapse of the bubble and so forms a complete thermodynamic cycle. For most purposes the overall temperature drop and associated losses may be neglected, but this is not always the case. For instance, when working near the critical point, or when comparing liquids with very different thermodynamic properties, the temperature effects may be important.

2.2 CAVITATION IN A FLOWING SYSTEM

So far, cavitation has been considered in terms of an individual bubble growing and collapsing. In a flowing system, however, the liquid is subject to local accelerations and at the points of highest velocity low pressures occur and cavities form; these cavities then extend until an area of lower velocity and higher pressures occurs when the cavities collapse. There is thus a succession of cavities forming and collapsing, which makes the bubble larger and apparently stable. The cavity still forms, of course, only when the low pressure is at or near the vapour pressure of the liquid.

The appearance of such cavitation is shown in Fig. 1, which also shows the corresponding pressure values through the system. As can be seen, the cavity collapse is unstable; the ends of the bubbles tend to break off and collapse, giving a pulsating end to the cavity. Where the cavities are symmetric on two surfaces, this leads to alternate collapse on the two surfaces, and vortex shedding behind the body.

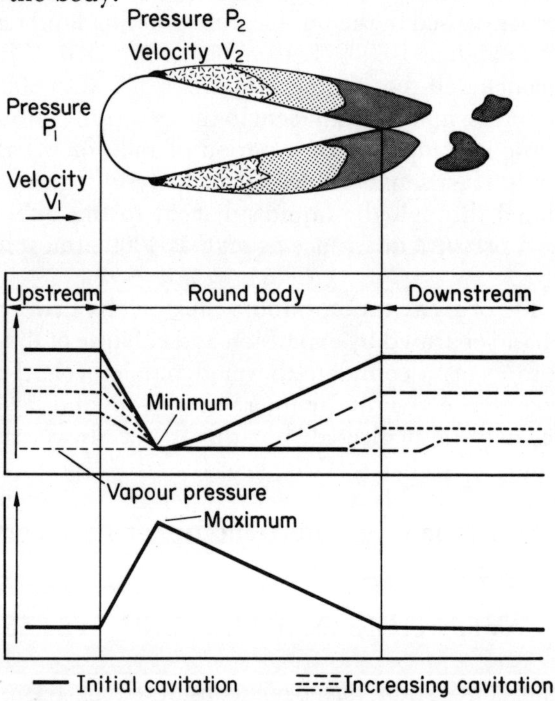

Fig. 1. Cavitation round a body in a fluid flow, such as hydrofoil, is shown diagramatically. The fluid flowing from left to right accelerates past the restriction caused by the body and reaches a point of maximum velocity and minimum pressure at the throat. If pressure upstream is sufficiently high the throat pressure will not fall below the vapour pressure, but as upstream pressure is reduced cavitation occurs and spreads rearwards.

The pressure and velocity relation in such a system obeys Bernoulli's equation, thus

$$\frac{P_1}{\rho} + \frac{V_1^2}{2} + gZ_1 = \frac{P_2}{\rho} + \frac{V_2^2}{2} + gZ_2 + \text{losses}$$

which for a horizontal system can be rearranged to give a cavitation coefficient representative of the system:

$$\sigma = \frac{P_1 - P_c}{\frac{\rho}{2} V_1^2}$$

where P_c is the critical pressure at point P_2 for which cavitation occurs; this is usually taken as the vapour pressure, although as explained above cavitation can occur at pressures either above or below this pressure.

The value of σ at which cavitation starts is usually called σ_{crit} and this is called the *inception point.* The *desinent point* is defined as the point where cavitation ceases with increasing pressure.

Because of the boundary layers (the pressure field around the body is non-uniform), the effect of nuclei, and the effect of geometry, the σ_{crit} is not a unique value. Thus it will vary accordingly from object to object and lead to air content, scale effects, etc. To a first approximation, however, it is a very useful correlation factor.

2.3 AIR CONTENTS EFFECTS

The presence of nuclei in the flowing stream means firstly that tensions will not be possible, and secondly if there are sufficient gas bubbles present cavitation will start by gas release at some pressure above the vapour pressure. (Also the cavity may be partly filled by gas at some pressure above the vapour pressure.) This means that cavitation will start at a σ value above the value corresponding to the vapour pressure. Thus, there will be an air content effect (Fig. 2). In addition, it is often found that with high air contents a hysteresis effect occurs on cavitation inception, and for this reason the desinent point which is less affected by hysteresis is usually used as the criterion of comparison.

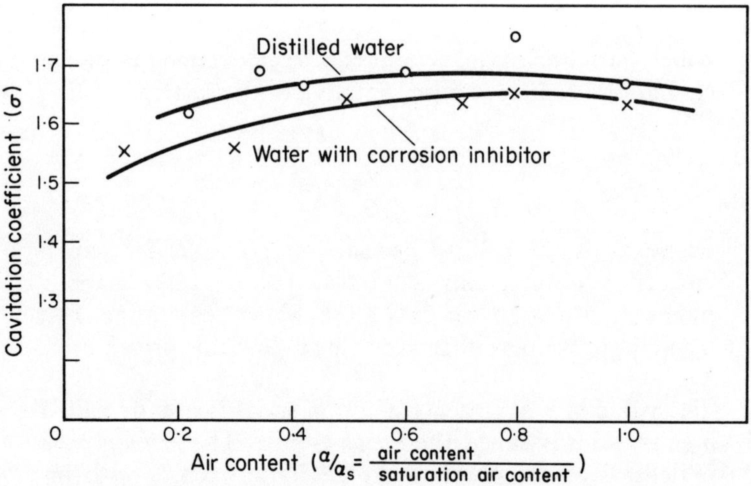

Fig. 2. Effect of air content on cavitation inception on a body (Fig. 1).

In the case of flow around static bodies such as cylinders, ogives and hydrofoils experimenters[5, 6] have found air content effects such as shown in Fig. 2, but on machines the position is less clear. In these cases, performance breakdown is usually more important than cavitation inception.

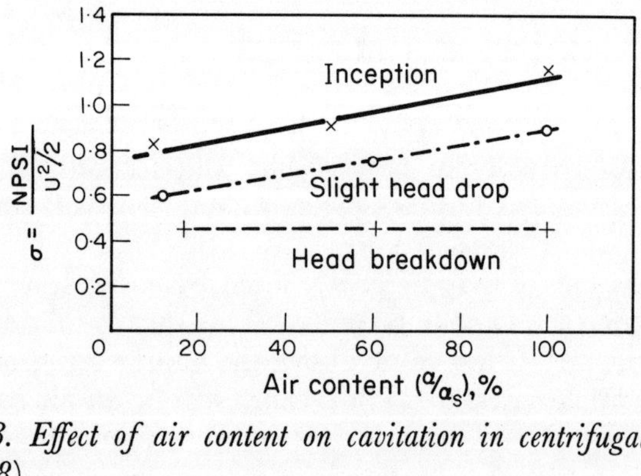

Fig. 3. Effect of air content on cavitation in centrifugal pump (Ref. 8).

Nevertheless, in all tests on machines—whether turbines, pumps or propellers—the effect of air is to increase the cavitation coefficient, although the rise varies from the negligible to the substantial (Fig. 3).(7–9)

2.4 SCALE EFFECTS

The time the fluid spends in the low pressure region will depend on the fluid velocity; hence at low velocities, when the fluid will be in the low pressure for a longer time than at higher velocities, it would be expected that cavitation would occur more easily. (The bubble has more time to grow.) It has been found experimentally that lower cavitation numbers can be attained at low velocities, and this is known as velocity scale effect.(11–12) Fig. 4 shows the velocity scale effect in the case of an ogive.

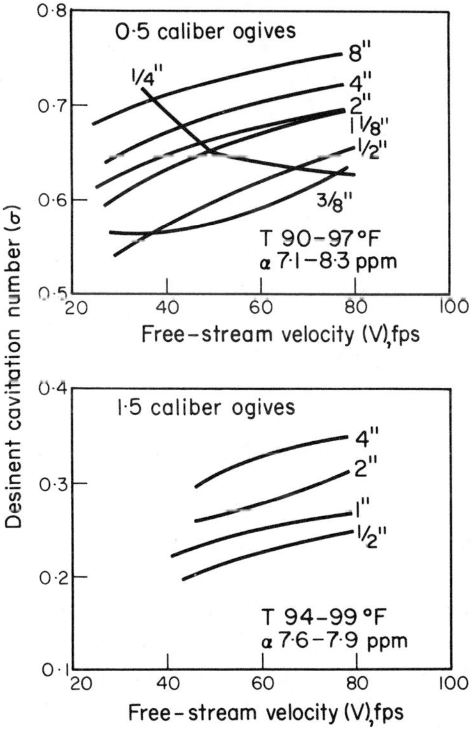

Fig. 4. Variation of σ with velocity and model diameter. Ogives in water (data from Ref. 10).

The scale effect of size does not correlate so well. Although most tests on static bodies show that σ increases with size,(11–13) tests on machines have usually shown that σ decreases with size. Similar correlations can be made on a basis of Reynolds numbers (Fig. 5) although, in fact, the Reynolds number is simply velocity $\times$ diameter as all the tests were carried out on water.

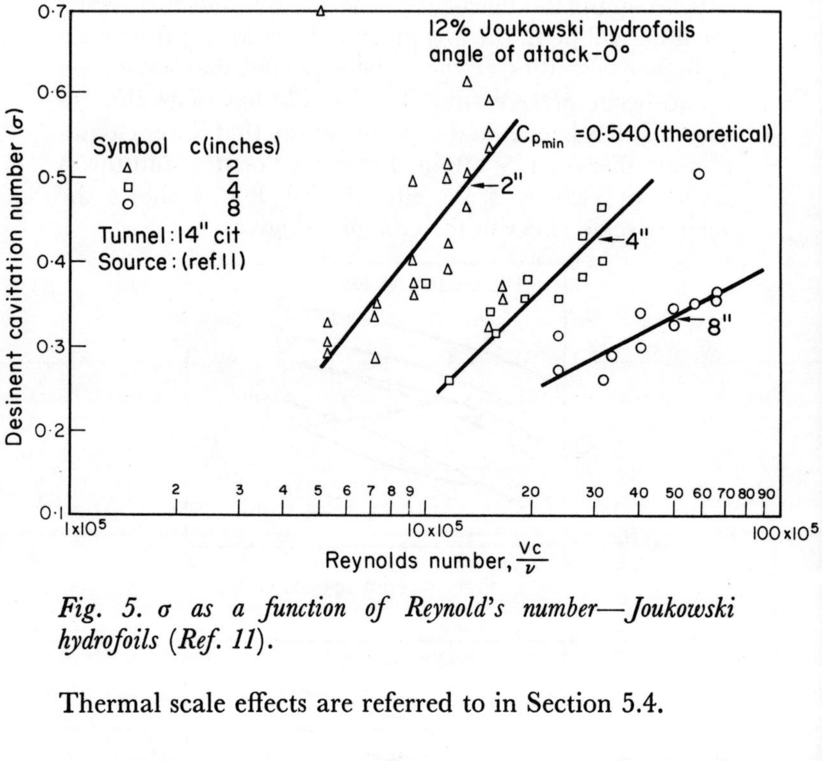

Fig. 5. σ as a function of Reynold's number—Joukowski hydrofoils (Ref. 11).

Thermal scale effects are referred to in Section 5.4.

2.5 NOISE

The collapse of cavitation bubbles creates shockwaves and hence noise. This is essentially "white noise" covering a wide frequency band; up to 1 MHz has been measured. There is some evidence to show that the smaller bubbles create the high frequency noise, the low frequencies being associated with the collapse of large bubbles.

Ultrasonic waves can be used to create cavitation. This technique is used in ultrasonic cleaning and in cavitation erosion testing equipment.

Noise can be used for detecting cavitation, and in general it gives the same inception (or desinent) point as visual methods, providing a less subjective assessment than visual means. Fig. 6 compares the use of visual methods and noise to detect the onset of cavitation on a Kaplan turbine.[14]

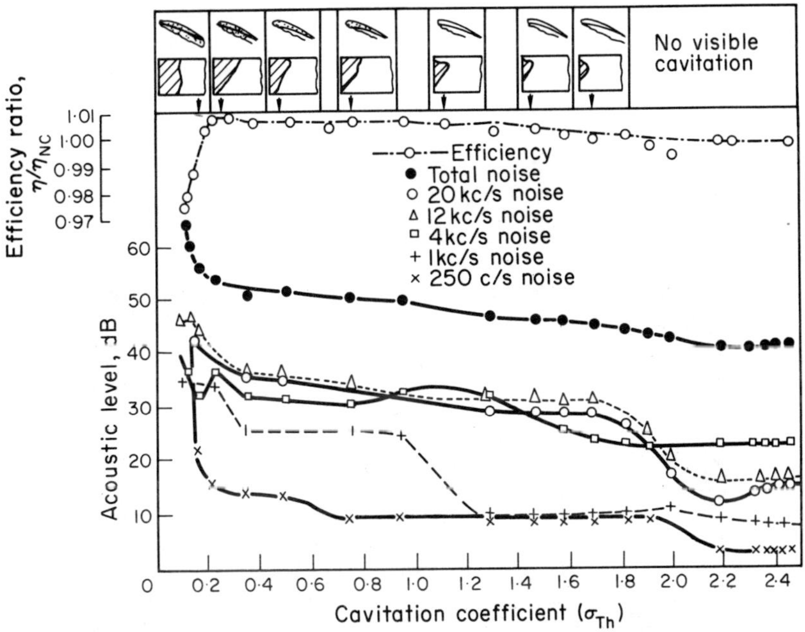

Fig. 6. Cavitation noise in Kaplan Turbine ($N_{st} = 400$).

Experiments have shown that there is a relationship between cavitation noise and erosion damage. Peak noise and peak erosion on a cylinder occur at the same time.[15] Development of this technique could be very useful for predicting when erosion occurs.

3. Cavitation Erosion

Erosion due to cavitation is a major problem. The high transient pressure and thermal shockwaves cause the material in proximity to the collapsing bubble to disintegrate, probably by a fatigue process. Pumps, turbines, propellers, as well as stationary devices like valves and dam spillways can suffer severe erosion (see Fig. 7). This may be so rapid as to require the replacement of a ship's propeller after a single voyage or a pump impeller after a few weeks. Turbine blades often require welding at regular intervals to repair erosion caused by cavitation. On the other hand, machines may run under cavitating conditions for years without suffering damage. The exact conditions of cavitation that lead to severe erosion are still unknown, but there is no doubt that erosion is accelerated in high velocity situations.

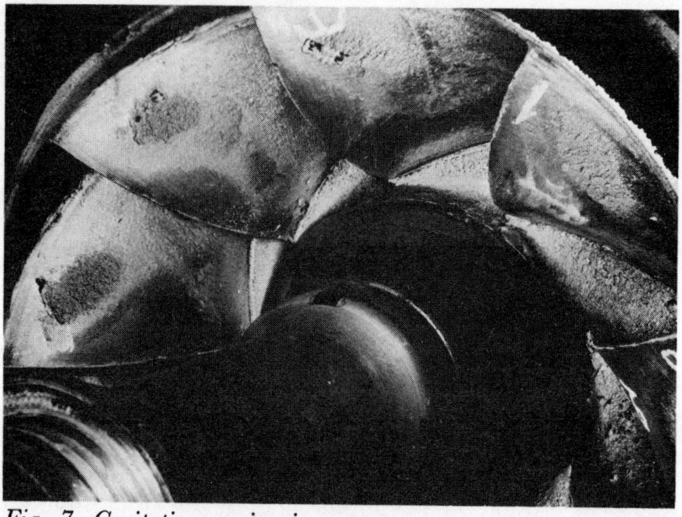

Fig. 7. Cavitation erosion in a pump.

3.1 CHOICE OF MATERIALS

Most materials when subjected to cavitation attack exhibit an initial incubation period during which no erosion occurs; this is shown as (A) in Fig. 8. The incubation period is followed by a period (B) marked by a rapid increase in erosion damage, then a fairly stable state (C) where the rate of erosion is mainly linear. Finally, as the surface becomes very pitted, the rate of erosion declines again (D).

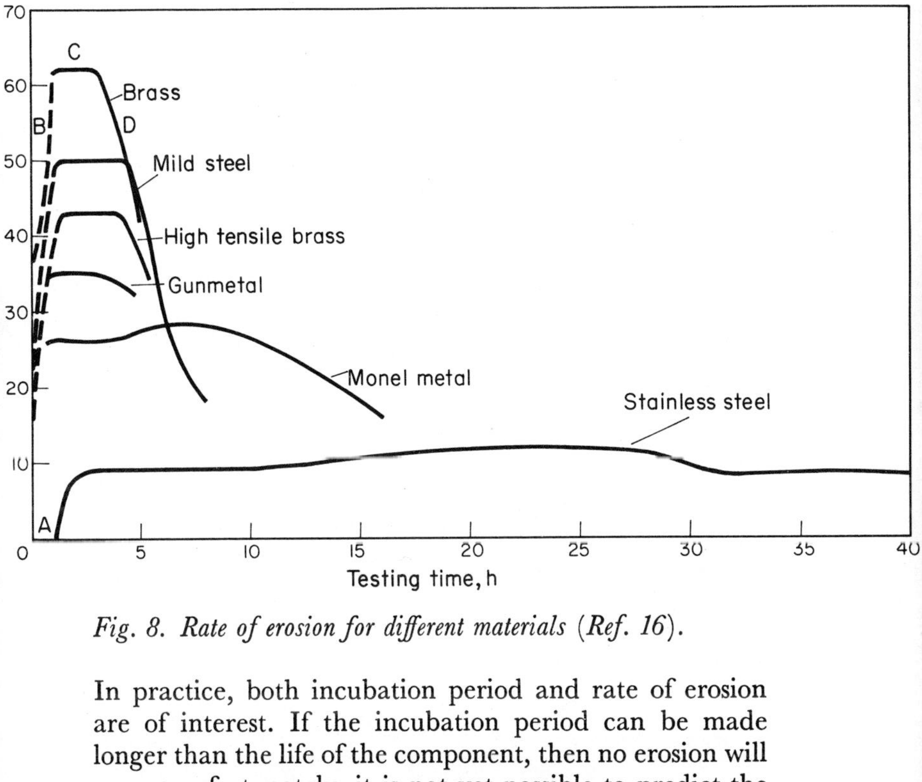

Fig. 8. Rate of erosion for different materials (Ref. 16).

In practice, both incubation period and rate of erosion are of interest. If the incubation period can be made longer than the life of the component, then no erosion will occur: unfortunately, it is not yet possible to predict the length of the incubation period for real components. By assessing the rates of erosion for different materials, however, it is possible to determine the relative resistances of these materials to erosion (Fig. 9). For this sort of test an accelerated method is commonly used.

Results of laboratory tests, see Table 1, indicate that aluminium bronze is one of the most erosion resistant materials in both fresh and sea water. Also, its corrosion

Table 1. Strength and Erosion Data for Engineering Metals

MATERIAL	RELATED B.S.	YOUNG'S MODULUS (ton/in^2)	TENSILE STRENGTH (ton/in^2)	ELONGATION (per cent)	HARDNESS HV	EROSION RATE (mm^3/h)
Mild steel	B.S.970 En1A	13600	38·5	16	200	5·5
Stainless steel	B.S.970 En56C	13660	51·2	26	253	1·3
Stainless steel	B.S.970 En57	13660	71·0	17	321	1·1
Stainless steel	B.S.970 En58B	13000	42·5	66	183	1·7
Stainless steel	B.S.970 En58J	13000	40·0	59	228	1·6
Stainless steel	B.S.1630 B	13600	44·4	26	228	1·7
Stainless steel	B.S.1630 B	13600	78·4	1	439	0·5
Stainless steel	B.S.1631 B Nb	13400	37·0	38	197	1·0
Stainless steel	B.S.1632 D	13400	36·4	48	176	1·3
Stainless steel	B.S.1648 C	13000	49·0	1	293	1·8
High tensile brass	B.S.1400 HTB–1	5900	25·9	17	120	4·7
Gunmetal	B.S.1400 G–1	5650	18·7	22	96	4·9
Aluminium bronze	B.S.1400 AB–1	7100	30·5	12	155	0·8
Aluminium bronze	B.S.1400 AB–2	8620	42·5	6	211	0·6
Novoston bronze	B.S.1400 CMA–1	8000	46·0	21	203	1·5
Monel K500(c.d.)		10700	54·0	19	289	2·8
Monel K500 (aged)		9400	73·0	13	396	1·2
Titanium	IMI 318A	7600	65·4	20	332	1·1
Titanium	IMI 680	7600	85·0	19	425	0·8
Ni-resist		8000	31·4	35	152	4·4
S.G. iron		11100	65·2	–	345	1·3

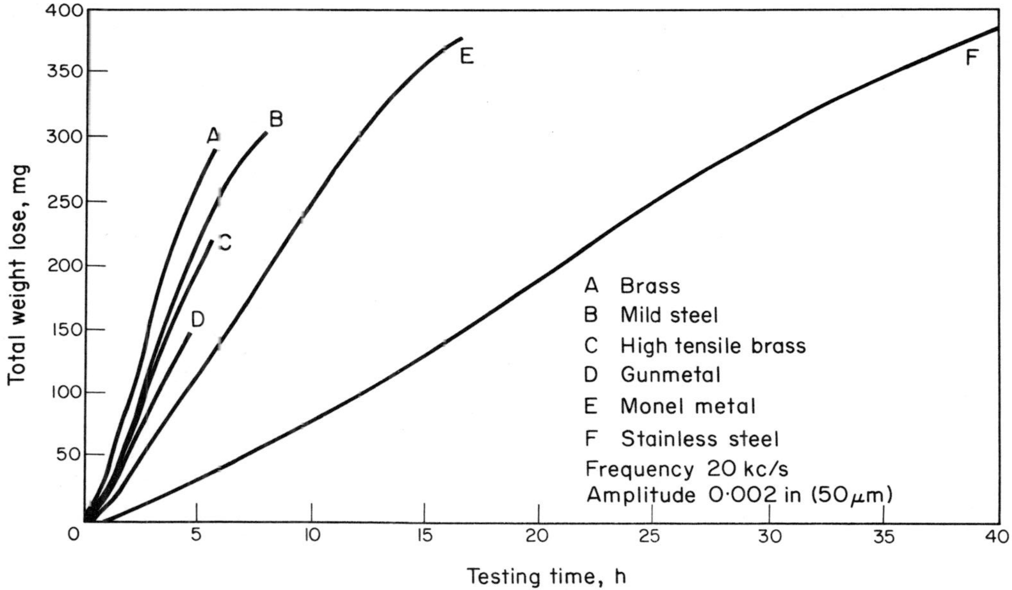

Fig. 9. Results of erosion tests on different materials (Ref. 16).

resistance is good. Aluminium bronze, however, is difficult to cast, and other alloys, such as Novoston, have been developed to overcome this difficulty. The corrosion and erosion resistances of these alloys are almost as good as those of aluminium bronze.

Stainless steels have a high erosion resistance, the best being precipitation hardening martensitic steels. Where corrosion is likely to present problems, the austenitic steels may be preferred, although their erosion resistance is inferior to that of the martensitic steels.

Nickel alloys, for example aged Monel metal, have good chemical and erosion resistance. Titanium has excellent erosion and corrosion resistance, but is rather expensive for ordinary applications.

Soft materials tend to deform without eroding. Lead, for example, behaves in this way, but it is of little practical use. Rubber and plastics have been used successfully as coatings (the erosion resistances of nylon and high impact polythene are similar to that of stainless steel). The main difficulty in using coatings is to secure good adhesion to the parent metal, particularly under the arduous conditions of cavitation collapse.

In corrosive fluids, damage is accentuated rapidly as a result of the dual effects of cavitation erosion and corrosion. Erosion damages the surface and removes protective films of oxide from the surface, giving ideal conditions for corrosion.

3.2 METHODS OF TESTING

Erosion can occur rapidly in practice, but for testing purposes the periods involved are too long, and it has been necessary to develop accelerated techniques. In all the tests used, erosion loss is ascertained by weight loss.

Vibratory Method

The disc-shaped specimen is mounted on the end of a

vibrator driven by a magnetostrictive transducer. The vibration is vertical and represents an approximate sine-wave of frequencies of about 20 kHz. The specimen is submerged in the liquid under test and the cavitation appears as small spherical bubbles on the surface of the specimen. Appreciable erosion on the specimen occurs in times of 30 minutes to 16 hours depending on the material.

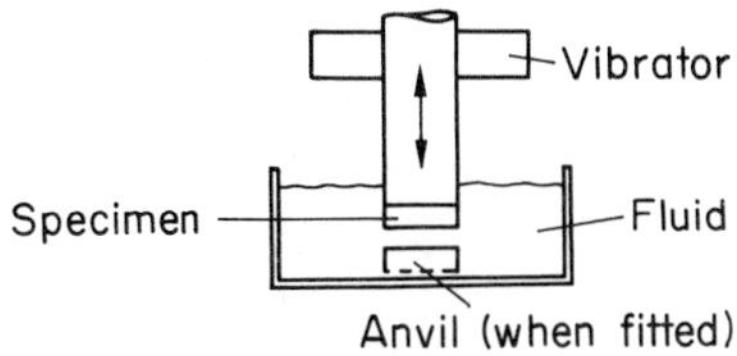

The cavitation cloud is affected by the viscosity and bulk modulus of the fluid. For fluids other than water, this method is altered in that the test specimen is mounted stationary below the vibrator and erosion is induced on this anvil.[17]

Impact Method

The specimens are mounted on the edge of a wheel which is rotated at high speeds and in impact with a jet of water. The result may not be true cavitation, but the test gives similar results to other techniques. It is useful especially for testing surface coatings and brittle materials. Appreciable erosion of even the hardest materials can occur in 4 minutes. The test is only really suitable for use with water, but it has also been used for investigating steam erosion in turbine blades. The latter, like raindrop erosion on aircraft, has a strong similarity to cavitation erosion.

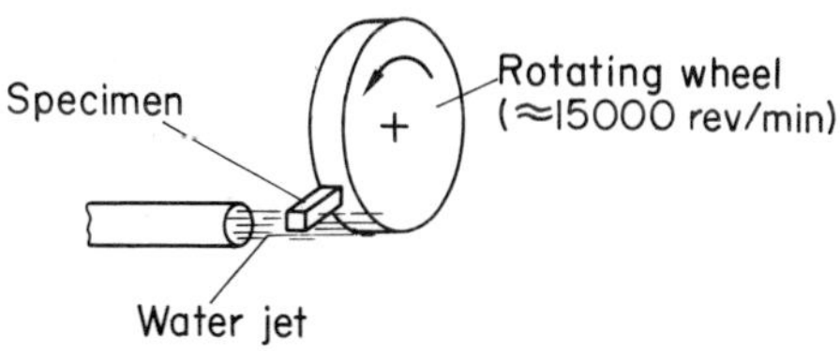

A similar, alternative method makes use of a rotating disc, on the surface of which the specimens are mounted.

Constricted Tube Water Tunnel

A constricted section in a water tunnel is used to obtain

high velocities. Cavitation forms either on a specimen mounted in the constriction, or on the wall of a venturi-type section. This method has the advantage that it is true hydraulic cavitation; on the other hand, a comparatively long time is needed (16–24 hours), to obtain erosion. The velocities must be greater than 80 ft/s if erosion is to occur.

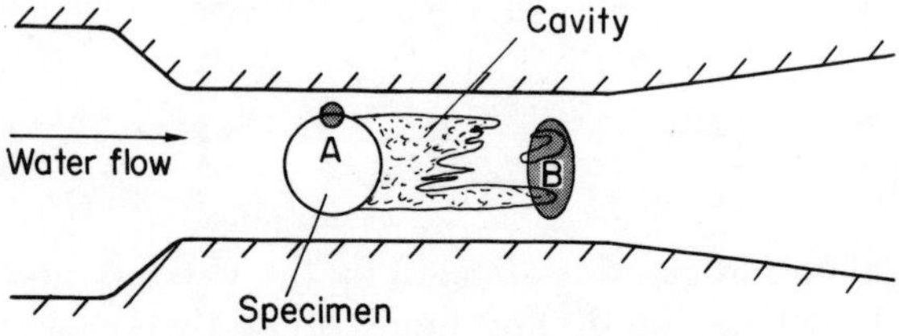

Comparison of Methods

On the whole, the different methods give comparable results. Fig. 10 compares results of the vibratory and water tunnel methods with those of the impact rig. The straight-line relationship shows that each method gives comparable results. The few exceptions include brittle materials or materials with a hard surface finish, which give false comparisons by one method.

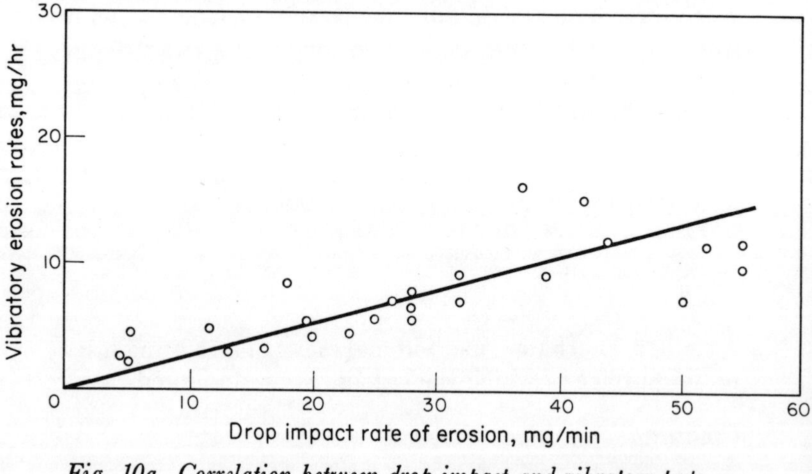

Fig. 10a. Correlation between drop impact and vibratory tests.

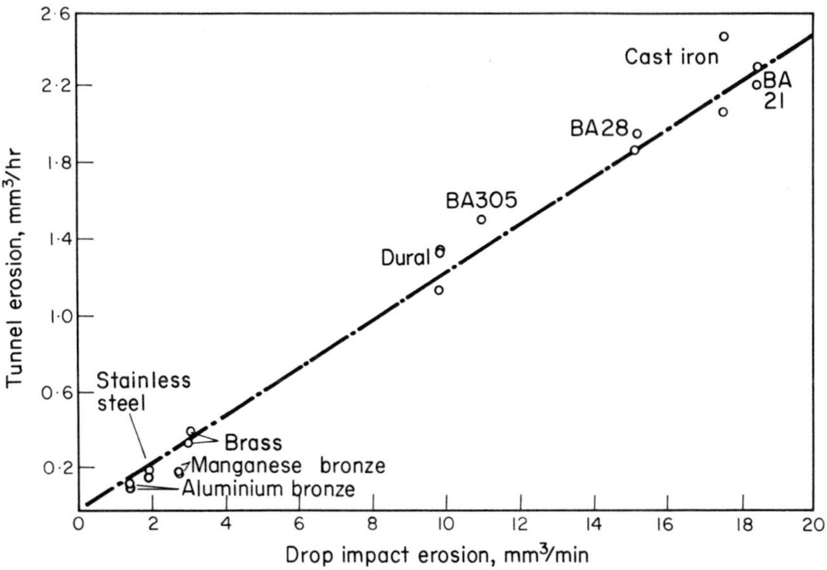

Fig. 10b. Correlation between drop impact and water tunnel tests.

3.3 CORRELATION OF EROSION RESISTANCE AND MATERIAL LIFE

Numerous attempts have been made to correlate the erosion resistance of materials with their physical properties. Strain energy, proof resilience, and ultimate resilience have all been considered with some success. None covers every case successfully, but it is clear that essential properties for good erosion resistance are high tensile strength, high hardness and good fatigue properties. Perhaps the best correlation is on the basis of hardness. Both erosion rate and incubation period are related to hardness. The greater the hardness the lower the erosion rate and the greater the incubation period. Figs. 11 and 12 show results obtained using the vibratory erosion method.

Erosion rate is also affected by velocity, pressure, temperature and air content. Constricted tube tests have shown that weight loss is proportional to V^7; drop impact tests show weight loss as being proportional to V^5. Erosion is approximately proportional to $(\text{pressure})^2$, and increased temperature appears to cause a small increase in erosion. Hobbs[18] found that erosion in water reached a maximum at 60°C.

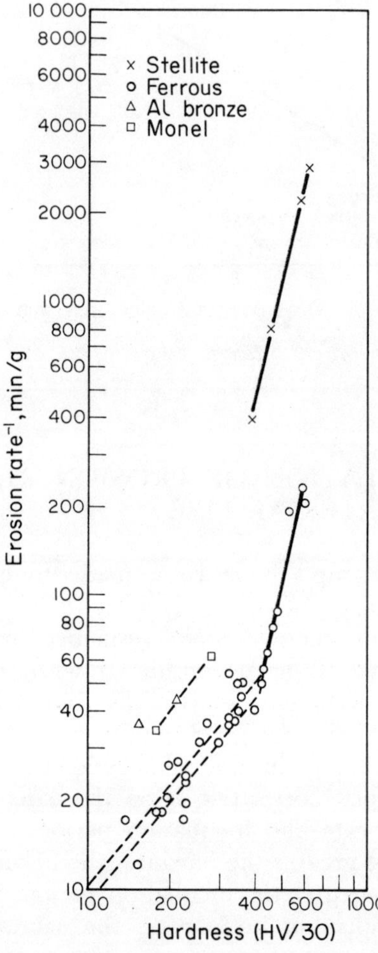

Fig. 11. Relation between erosion rate and hardness.

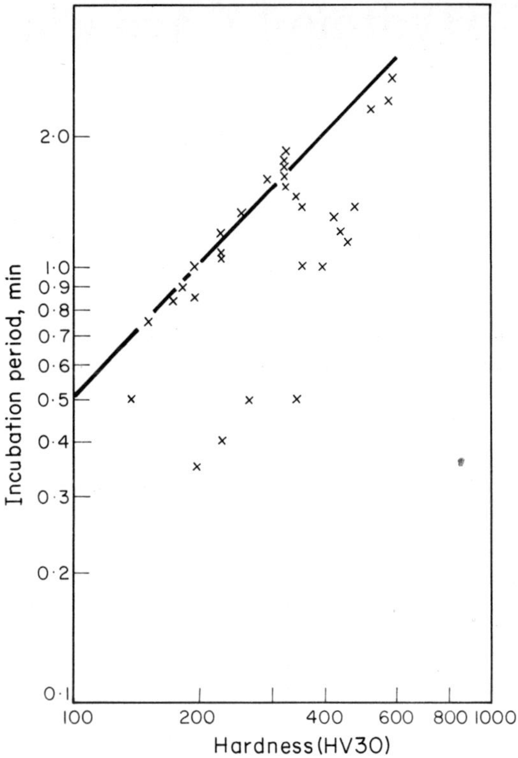

Fig. 12. Relation between incubation period and hardness.

4. Hydrofoil Cavitation

4.1 CONVENTIONAL HYDROFOILS

Cavitation causes loss of lift and increased drag on hydrofoils such as those used for hydrofoil boats, stabilisers, propellers, pumps and turbines. These effects are usually determined from water tunnel tests, although the inception of cavitation can be estimated from pressure distributions determined by theory or by air tests.[19]

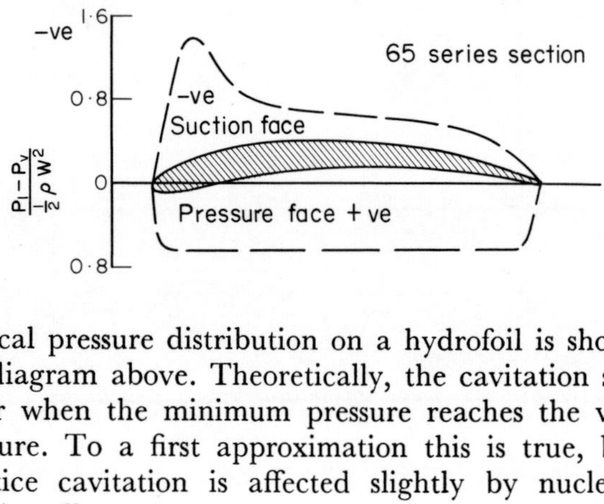

Typical pressure distribution on a hydrofoil is shown in the diagram above. Theoretically, the cavitation should occur when the minimum pressure reaches the vapour pressure. To a first approximation this is true, but in practice cavitation is affected slightly by nuclei and velocity effects.

The cavitation performance can be expressed in terms of the cavitation coefficient $\sigma_b = (P_1 - P_v)/\rho W_1^2/2$.

After inception, as the cavity spreads, the circulation and hence pressure distribution are affected until eventually the lift drops and the drag increases. Fig. 13 shows the effect on the lift of a hydrofoil; Fig. 14 shows how the lift/drag ratio is reduced.[20]. Cavitation commences earlier at high incidences, so that for good cavitation performance incidences should be low, less than 3°. Thin hydrofoils with little camber also have good cavitation performance. Unfortunately, these requirements for good cavitation performance do not necessarily make for good overall performance of a machine, and compromise is usually necessary.

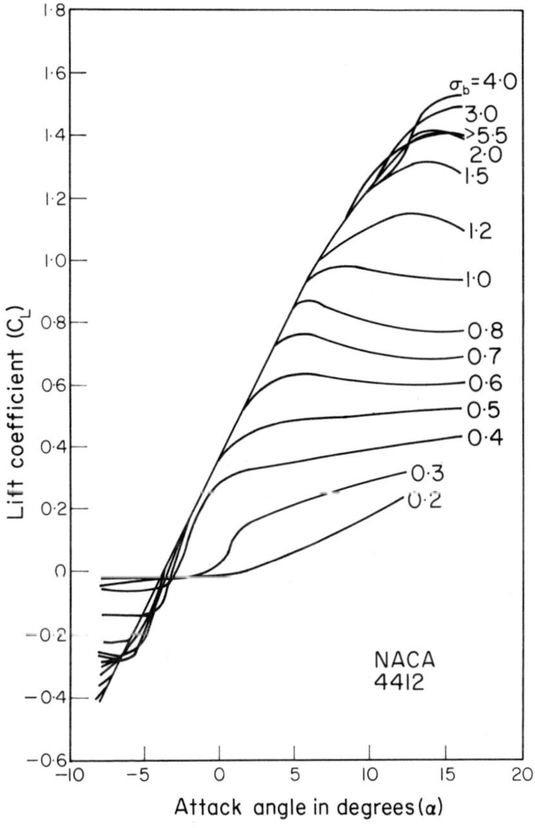

Fig. 13. Lift coefficient as a function of angle of attack and cavitation number—NACA 4412 Hydrofoil (Ref. 20).

The extent of cavitation is dependent on cavitation number and angle of incidence; by the time the cavity becomes greater than the chord length, the lift of a hydrofoil has dropped significantly and L/D is low.

Fig. 15 shows the complete characteristics of a hydrofoil.

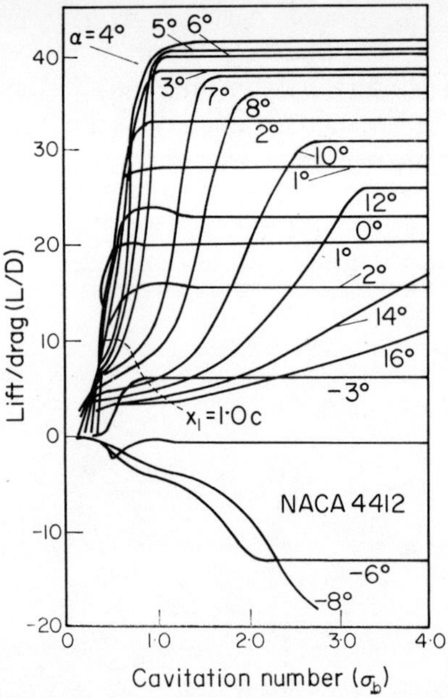

Fig. 14. Lift/drag ratio as a function of cavitation number and angle of attack for the NACA 4412 hydrofoil. The dashed line at the lower left indicates the cavitation number at which the cavitation extends downstream to the trailing edge of the hydrofoil.

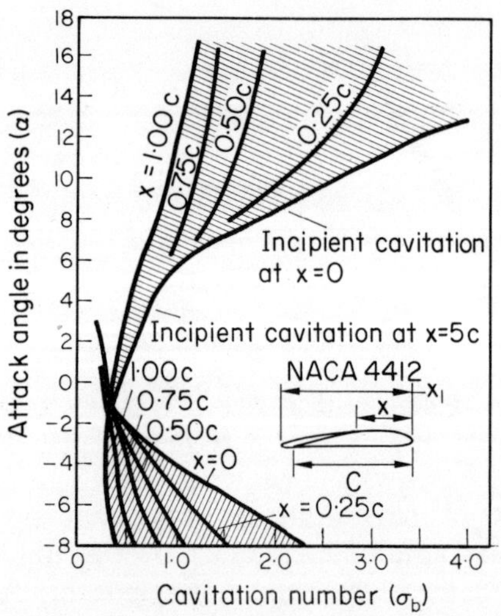

Fig. 15. Cavitation diagram for the NACA hydrofoil. The shaded areas are regions of cavitation number and attack angle wherein the cavitation collapses on the hydrofoil (Ref. 20).

4.2 CASCADES OF HYDROFOILS

When a series of hydrofoils is used in "cascade" together, as in a propeller or pump impeller, each hydrofoil has an effect on its neighbour. These cascade effects alter the cavitation characteristics as well as the lift and drag. Generally, the effect is to decrease suction pressure peaks, which also decreases lift for a given incidence. It therefore has a beneficial effect on cavitation inception. Experimental data relating to cavitation tests on cascades of hydrofoils are few.[21, 22]

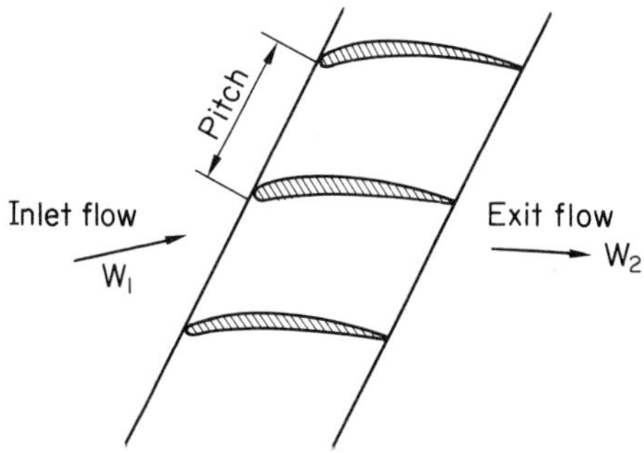

4.3 SUPERCAVITATING HYDROFOILS

There are certain cases where velocities are so high that cavitation cannot be avoided; they include, for example, high-speed boats, their propellers and high-speed pumps. For these cases special supercavitating hydrofoils have been developed. By using a sharp leading edge a cavity is formed over the complete suction side; it collapses downstream and erosion damage is avoided (Fig. 16). By using

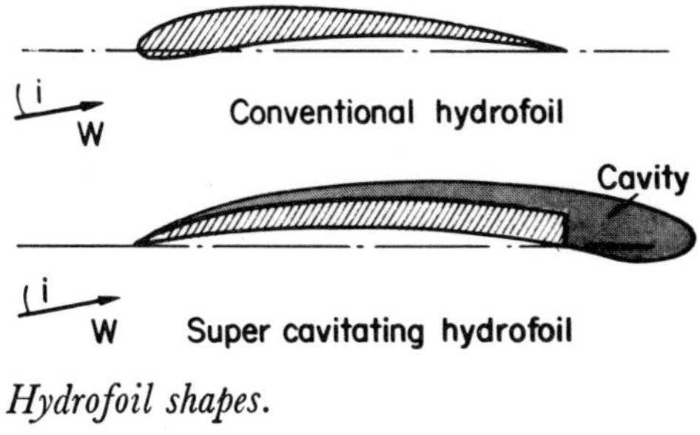

Fig. 16. Hydrofoil shapes.

these methods, reasonable lift and drag values are attained; they are better than those obtained using conventional sections, but not so good as those using non-cavitating sections.[23] Similar cascade effects are evident as in the case of conventional sections. There is a fair amount of data on isolated supercavitating sections, but there is only a very limited amount of experimental data on cascade supercavitating sections.[22] There are, however, theoretical data.[24, 25]

5. Cavitation in Pumps

5.1 DEFINITIONS AND PARAMETERS

Pumps cavitate when the suction pressure drops too low or when the speed is too high. The cavitation occurs when

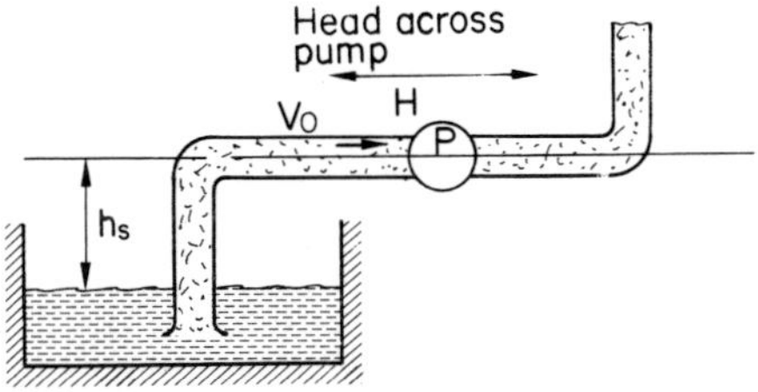

the local pressure on the suction side of the blade drops to near vapour pressure, as in the case of hydrofoils. This occurs when

$$\frac{P_a}{\rho} - gh_s - g\Delta h = \frac{P_v}{\rho}$$

where P_a is atmospheric pressure, P_v is vapour (or critical) pressure, where $g\Delta h$ can alternatively be expressed as

$$\frac{V_{m1}^2}{2} + \lambda \frac{W_1^2}{2}$$

where λ is an empirical coefficient based on blade shape (it is equal to σ_b the hydrofoil cavitation coefficient).

Thoma[55] suggested that Δh is proportional to H, and defined a cavitation coefficient

$$\sigma_{Th} = \frac{\Delta h}{H}$$

where H is the total head across the machine, Δh is known as the Net Positive Suction Head (*NPSH*), so

$$NPSH = \frac{P_a}{\rho g} - h_s - \frac{P_v}{\rho g}$$

or in SI units the Net Positive Suction Energy (*NPSE*)

$$NPSE = gNPSH = \frac{P_a}{\rho} - gh_s - \frac{P_v}{\rho}$$

The *NPSE* thus equals the total energy at the suction flange less the vapour energy. (Note: The losses in the suction pipe must be subtracted from this if they are significant.)
Thus

$$\sigma_{Th} = \frac{NPSH}{H} \quad \text{or} \quad \frac{NPSE}{Y}$$

where Y = total specific energy = gH.
We may also redefine blade cavitation coefficient

$$\sigma_b = \frac{P_1 - P_v}{\frac{1}{2}\rho W_1^2}$$

which is the same as the hydrofoil coefficient. P_1 is static pressure in front of the blade and W_1 the relative inlet

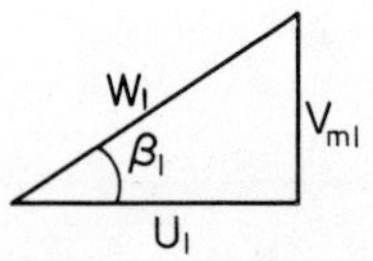

Inlet velocity triangle

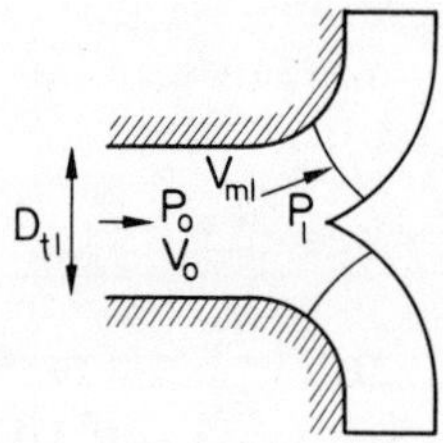

velocity. Another useful parameter is suction specific speed

$$S = \frac{N\sqrt{Q}}{NPSH^{3/4}}$$

This has different values according to the units. In British systems N is rev/min, Q is Imp. gal/min and *NPSH* is in ft. It is preferable therefore to use the non-dimensional form

$$K_s = \frac{\omega\sqrt{Q}}{(NPSE)^{3/4}} \quad \text{or} \quad \frac{\omega\sqrt{Q}}{(gNPSH)^{3/4}}$$

where ω is rad/s and the other units are consistent (i.e. in SI, Q is in m^3/s, *NPSE* is in J/kg and *NPSH* is in m). Note that S (British units) = 2495 K_s.
σ_{Th} is related to S by

$$\sigma_{Th}^{3/4} = \frac{N_s}{S} = \frac{K_n}{K_s}$$

where N_s is specific speed and K_n is non-dimensional specific speed. For pumps geometrically similar at different speeds or of different diameters then the scaling laws can be used:

$$\frac{NPSH_1}{NPSH_2} = \left(\frac{N_1}{N_2}\right)^2 \left(\frac{D_1}{D_2}\right)^2.$$

5.2 EFFECT ON MACHINES

Although it is possible to predict approximately when cavitation inception occurs in a pump, in most applications it is the point where cavitation affects performance that is of interest. By this time, the inception cavity has grown into a sizable cavity, and it is necessary to determine the size of this cavity as well as its effect on the machine. There are also the complicating factors of nucleation, boundary layers, gas contents, and secondary flows. Thus it is not possible to be exact, and most estimations of these factors have been made by empirical methods.

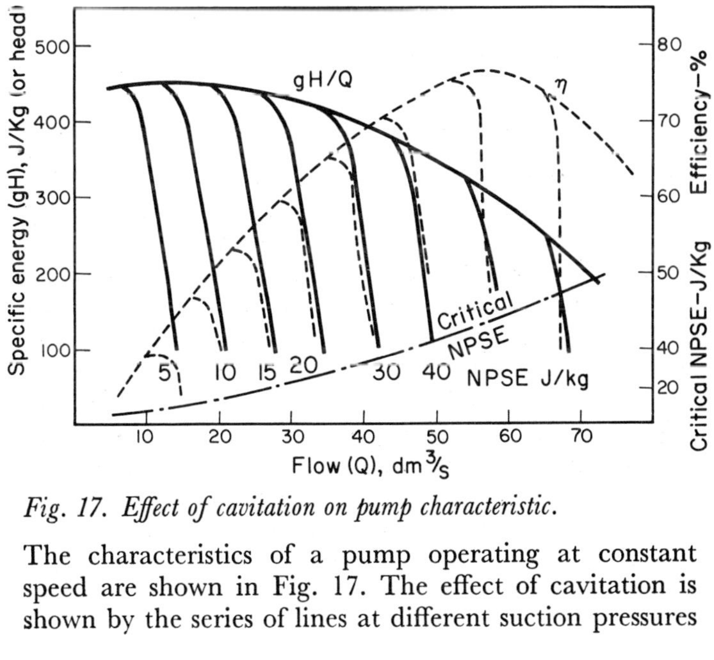

Fig. 17. Effect of cavitation on pump characteristic.

The characteristics of a pump operating at constant speed are shown in Fig. 17. The effect of cavitation is shown by the series of lines at different suction pressures or *NPSE*. Thus as the *NPSE* is reduced, the "fall off" in performance occurs at diminishing flows. It is also apparent that cavitation affects head and efficiency in the same manner.

If we define a critical *NPSE* as the point where the head falls by 2 per cent from the non-cavitating characteristic (it could also be 1 per cent, 5 per cent or 10 per cent), then we can plot these critical *NPSE* (or *NPSH*) values against flow (Fig. 17). (Note: *NPSE* (*NPSH*) is often confused with critical *NPSE* (critical *NPSH*).)

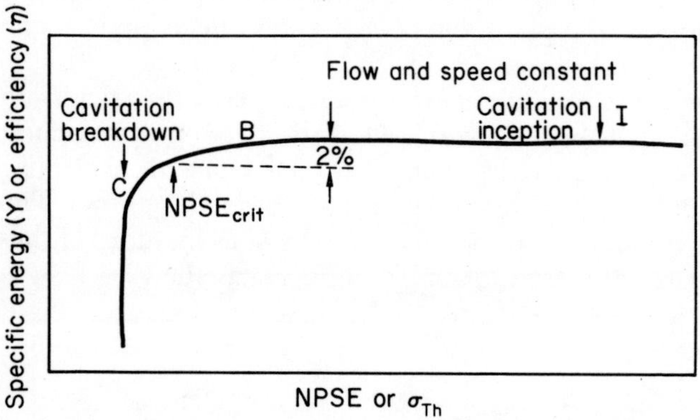

Fig. 18. Cavitation characteristic of a pump.

The cavitation characteristic is often presented in a different manner. For a constant speed and flow (possibly best efficiency point), specific energy (head) or efficiency is plotted against *NPSE* (*NPSH*) or σ_{Th} as in Fig. 18. This shows clearly the wide difference in *NPSE* (*NPSH*) between inception and performance breakdown (one may be 3 or 4 times the other). In a practical application, this would result in the pump running at a speed of $\frac{1}{2}$ to $\frac{1}{3}$ to avoid cavitation altogether, as opposed to running with cavitation just above the "breakdown" point. This would mean a much larger and more expensive pump to satisfy the condition of no cavitation, and for this reason it is rarely done in commercial practice. However, the disadvantage of running with cavitation is the risk of erosion damage.

Centrifugal pumps (and reaction turbines) both exhibit this sudden drop in performance due to choking of the impeller passages. Axial pumps (and turbines) tend to have a much more gradual fall in performance.

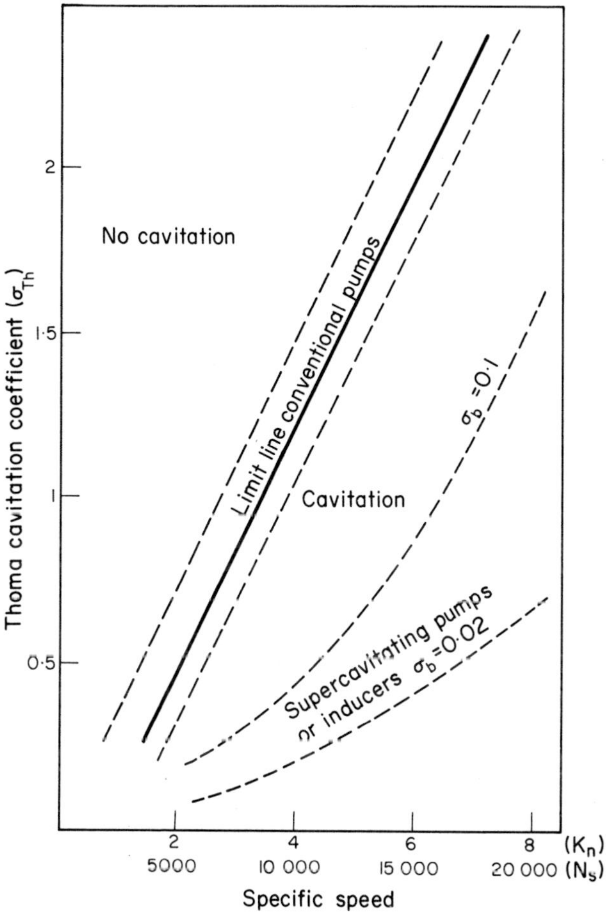

Fig. 19. Cavitation limits for pumps.

The results of many tests on pumps have been correlated by the Hydraulic Institute in the USA.[26] It has been possible to construct a graph relating σ_{Th} to specific speed K_n or N_s, for critical and other conditions (Fig. 19). This type of diagram is useful in classifying pumps on the basis of good and bad design. It can provide information which is extremely useful for comparing a particular design or for estimating the required *NPSE* (*NPSH*) for a new plant or installation.

It will also be apparent that as $\sigma_{Th}^{3/4} = K_n/K_s$ the slope of these lines corresponds to constant suction specific speed. Thus $K_s = 2{\cdot}8$ ($S = 7100$) corresponds to limit-values for ordinary pumps. Pumps may operate safely below this value, although the possibility of erosion should be borne in mind. It is not safe to operate at greater suction specific speeds.

For multistage and double entry pumps, only the *NPSE*, Q and N relative to a single impeller should be used.

5.3 OPTIMUM DESIGN

With reference to the blade inlet drawing and velocity triangle (see p. 40), applying Bernoulli's equation to stations 0 and 1

$$\frac{P_0}{\rho} + \frac{V_0^2}{2} = \frac{P_1}{\rho} + \frac{V_{m1}^2}{2} + \text{loss}$$

for a horizontal pump, or

$$NPSE = \sigma_b \frac{W_1^2}{2} + \frac{V_{m1}^2}{2} + \text{loss.}$$

It can be shown that loss $\approx 0{\cdot}04(V_{m1}^2/2)$, so

$$NPSE = \frac{V_{m1}^2}{2}(1{\cdot}04 + \sigma_b) + \frac{\sigma_b U_1^2}{2}.$$

or alternatively

$$= k_1 \frac{V_{m1}^2}{2} + k_2 \frac{U_1^2}{2}$$

where k_1 and k_2 are empirical coefficients, so

$$K_s = \frac{2{\cdot}98\sqrt{(1-\lambda^2)}}{\phi\left((1+\sigma_b) + \dfrac{\sigma_b}{\phi^2}\right)^{3/4}}$$

where σ_b and ϕ are both tip values.

To find the condition for maximum suction specific speed (or minimum *NPSE*), differentiate with reference to ϕ

hence $\quad \phi = \dfrac{V_{m1}}{U_1} = \sqrt{\left\{\dfrac{\sigma_b}{2(1+\sigma_b)}\right\}} = \sqrt{\left(\dfrac{k_2}{2k_1}\right)}$

and $\quad \dfrac{NPSE}{U^2/2} = \tfrac{3}{2}\sigma_b,$

hence the optimum diameter

$$D_{t1} = 1{\cdot}37\left[\sqrt{\left\{\frac{2(1+\sigma_b)}{\sigma_b}\right\}}\left(\frac{Q}{\omega}\right)\frac{1}{(1-\lambda^2)}\right]^{\frac{1}{3}} \qquad (1)$$

where σ_b is related to tip values, or

$$D_{t1} = 1{\cdot}72\left[\sqrt{\left\{\frac{2(1+\sigma_b)}{\sigma_b}\right\}}\left(\frac{Q}{\omega}\right)\frac{1}{(1-\lambda^2)(1+\lambda)}\right]^{\frac{1}{3}} \qquad (2)$$

where σ_b is at mean radius. Note: If empirical coefficients are used then

$$\left\{\frac{2(1+\sigma_b)}{\sigma_b}\right\}$$

should be replaced by

$$\sqrt{\left(\frac{2k_1}{k_2}\right)}.$$

This equation is shown in graphical form in Fig. 20. It will be noted that the greater is Q/ω the greater is the diameter, and also the lower is σ_b the greater is the diameter.

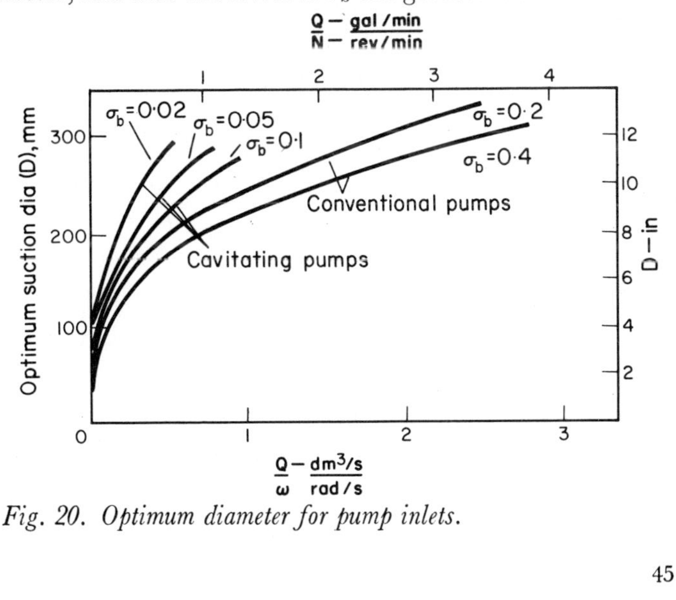

Fig. 20. Optimum diameter for pump inlets.

In other words, a pump with good cavitation performance will need a larger inlet diameter than ordinary pumps.

By rearranging these equations other relationships can be attained:

$$D_{t1}^2 = \frac{4Q}{\pi(1-\lambda^2)}\left\{\frac{3}{2}\frac{(1+\sigma_b)}{NPSE}\right\}^{1/2}$$

or

$$N = 35{\cdot}2\sqrt{\left(\frac{1-\lambda}{1+\lambda}\right)}\frac{(NPSE)^{3/4}}{\sqrt{Q}}\left\{\frac{1}{(1+\sigma_b)\sigma_b^2}\right\}^{1/4}$$

For design purposes Q and $NPSH$ are usually specified and hence D_{t1} and N can be determined for optimum cavitation performance. λ should be kept as low as possible, approximately equal to 0·3.

The cavitation coefficient σ_b is determined by blade shape and incidence. For inception conditions it can be determined from cascade data,(19) using the equations for tip conditions (1) Fig. 21. If the incidence is low (about 3 or 4°) and the blades are thin and well shaped, $0_b \approx 0{\cdot}25$ can be achieved without cavitation. Empirical data by Gongwer(27) suggests $k_1 = 1{\cdot}80$ and $k_2 = 0{\cdot}23$.

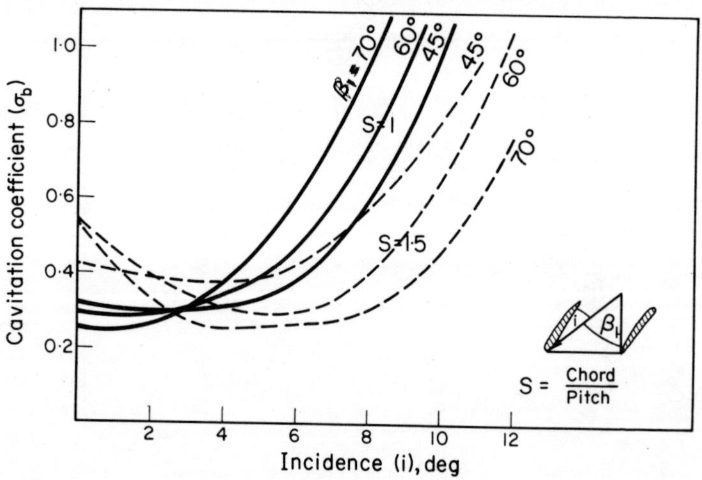

Fig. 21. Cavitation inception characteristics NACA-65-010 hydrofoil.

For performance breakdown, lower limiting σ_b values can be obtained from Fig. 22 (derived by theoretical methods[27] and a minimum σ_b of 0·08 is obtainable with low incidences and thin blades. Alternatively, using empirical methods $k_1 = 1{\cdot}41$ and $k_2 = 0{\cdot}085$, which gives approximately equivalent values.

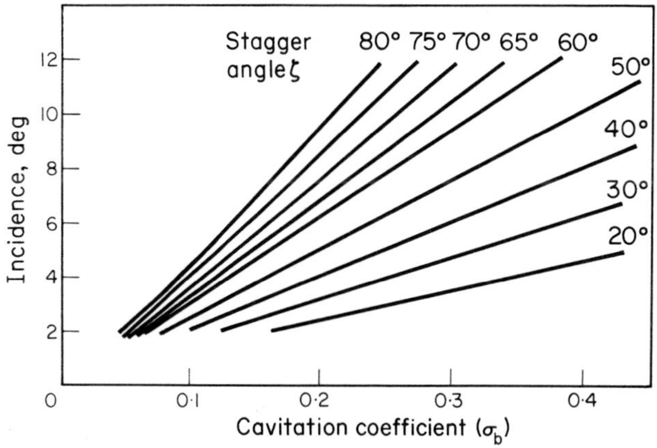

Fig. 22. Cavitation breakdown limits.

Although these methods give the optimum performance, it is possible to operate slightly off the optimum and still get an acceptable performance. It is often useful to plot *NPSH* against D_{t1} to show the effect of these variations.

5.4 THERMODYNAMIC EFFECTS

The cavitation performance of pumps in liquids other than water or in water at high temperatures is better than in cold water (Fig. 23).[28, 29] This can be explained in that the performance of the pump is affected by the size of cavity. At higher vapour pressures a higher depression $\Delta NPSE$ is required to produce the same volume of bubbles at the same head loss. Thus it will require a larger amount of work and hence a greater temperature increment Δt.

Consider the two cases of a liquid at 70 and 300 °F (or alternatively two liquids) (Fig. 24). The saturation temperature corresponding to the reduced pressure in the low pressure region of the impeller would be lower by an amount Δt. If sufficient time is allowed Δh_f J/kg of liquid passes through the low pressure zone and will be available for vaporisation of the liquid. Δh_f is change of enthalpy.

$$\Delta h_f = C_P \Delta t$$

where C_P is specific heat.

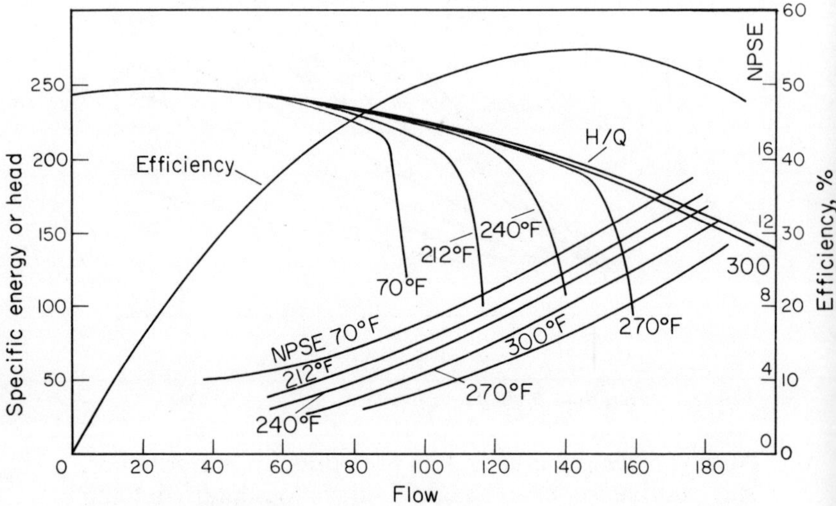

Fig. 23. Effect of temperature on cavitation in a pump (Ref. 28).

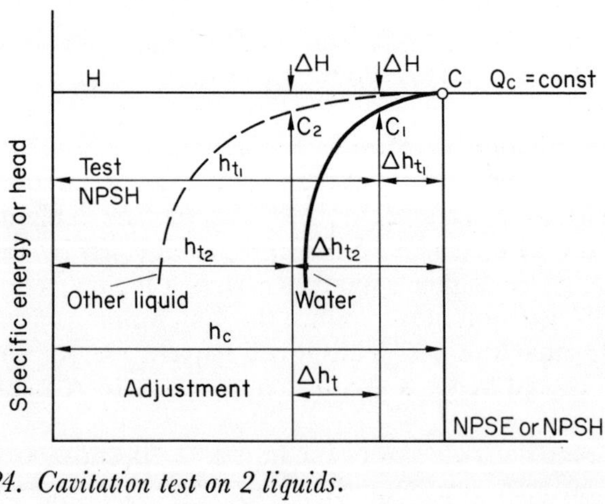

Fig. 24. Cavitation test on 2 liquids.

If thermal equilibrium is restored, then per unit mass:

$$\Delta h_f = \frac{\bar{V}_v}{v_v} L$$

$\bar{V}$ is volume,
v is specific volume,
L is latent heat,
subscript v represents vapour, and
subscript L represents liquid, so

$$\frac{\bar{V}_v}{\bar{V}_L} = \frac{v_v}{v_L} \frac{\Delta h_f}{L} = B.$$

From the Clapeyron equation

$$\Delta h_f = \frac{\Delta NPSH\ T\, v_v\, C_P}{L\, v_L}$$

hence

$$B = \frac{\Delta NPSH\, C_P\, T}{J' L^2} \left(\frac{v_v}{v_L}\right)^2$$

or

$$B_1 = \frac{B}{\Delta NPSH} = \frac{C_P T}{J' L^2} \left(\frac{v_v}{v_L}\right)^2$$

where T is temperature,
J' is mechanical equivalent of heat.

Stepanoff[28] showed that the relation of B_1 to the vapour

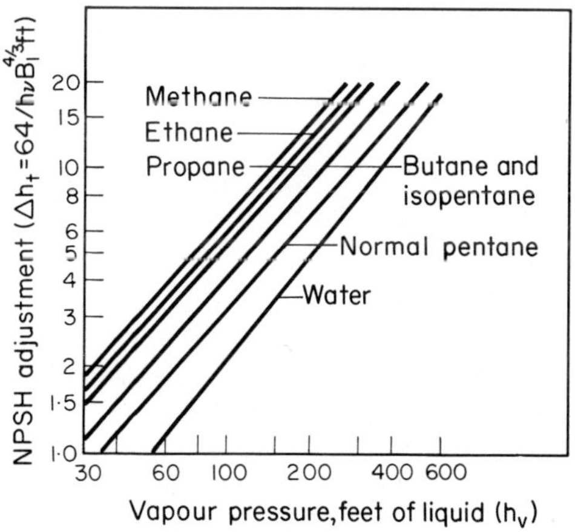

Fig. 25. NPSH adjustment for hydrocarbons.

pressure h_v is a straight line of slope 37° on log/log scales, and hence the empirical relation

$$\Delta NPSH = \frac{64}{h_v B_1^{4/3}} \quad \text{(ft units)}$$

Fig. 25 shows the effect for some common chemicals. Ward and Sutton[29] correlated a number of test results and concluded that

$$\Delta NPSH = \frac{r_v}{B_1}$$

where r_v = ratio of volume of liquid and vapour, which in practice they showed was ≈ 0·8.

Hence $$\Delta NPSH \approx \frac{0{\cdot}8}{B_1} \quad \text{(ft units)}$$

which is probably sufficiently accurate for most estimation purposes when it is remembered that $\Delta NPSH$ is small anyway.

5.5 SUPERCAVITATING PUMPS AND INDUCERS

There are cases where, for technical or economic reasons, the cavitation limitations imposed by ordinary pumps are unacceptable. Examples include applications where weight is important, such as aircraft or rocket pumps, and where high speed pumps are needed to minimise weight or where the liquid is near its vapour pressure as in chemical plant, extraction pumps or boiler feed pumps. There are also economic advantages to be gained by running ordinary pumps (particularly high pressure pumps) faster. In all cases, improved cavitation performance is desirable.

The achievement of better cavitation performance means firstly running at optimum conditions (Section 5.3) and secondly lower cavitation numbers, σ_b. The latter can be attained by running the blade with a lot of cavitation and by having few blades and a sharp edged section. By this means, σ_b of 0·04 or lower can be obtained.

There are two such types of pumps; inducers [30] and supercavitating pumps (Fig. 26).[27, 31] The former has long blades and the cavity formed at the leading edge collapses on the blade. The latter has short blades and the cavity collapses behind the blades; for this reason it is less susceptible to erosion than the inducer. Both types are axial pumps and the pressure produced is therefore small.

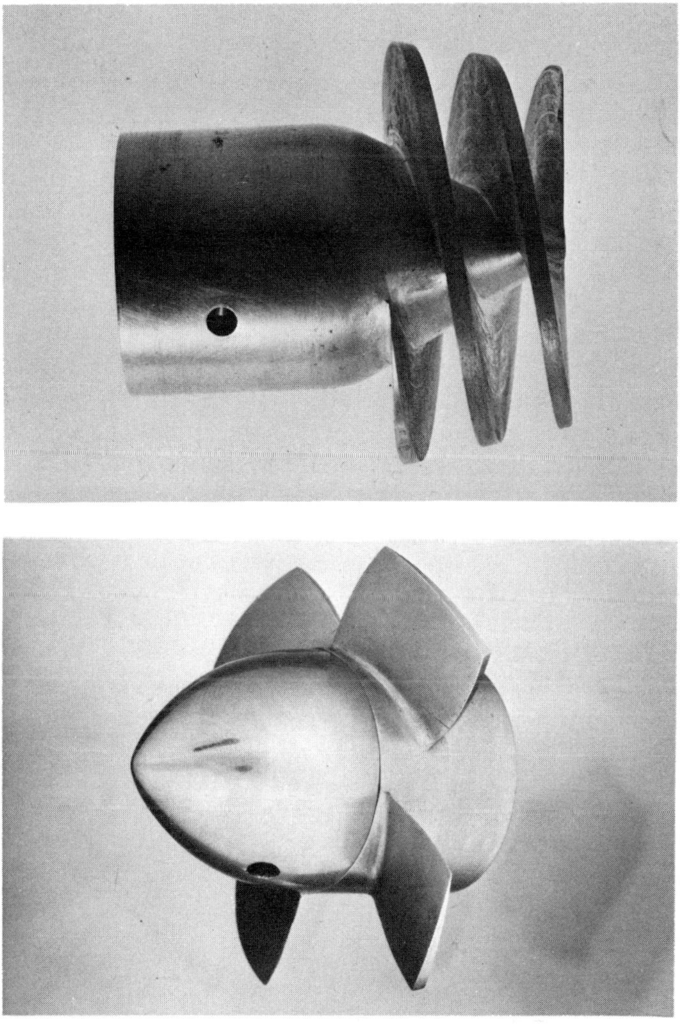

Fig. 26. Inducer (top) and supercavitating pump (bottom).

For many applications they would need to be staged as a suction booster to a centrifugal pump, to achieve the low *NPSE* (*NPSH*) and required overall pressures. In these cases, because σ_b is lower than that of the centrifugal pump, the optimum diameter is greater; thus, the cavitating stage is bigger than the inlet of the centrifugal stage that follows it.

By using such inducers or supercavitating pumps, suction specific speeds of 8 (20,000) can easily be attained; by introducing added refinements, speeds of 12 to 16 may be reached.

5.6 EROSION IN PUMPS

The basic considerations are as outlined in Section 3. If the pump does not cavitate, there will be no erosion. Unfortunately, as already stated, this means in many cases that the pump must run so slowly and it must be so big as to be uncompetitive. Thus, many pumps run in the area B to I, Fig. 18, with some cavitation and erosion likely. At the present state of knowledge, it is not possible to predict whether a particular installation will erode or not. However, it is possible to make some general comments. It is necessary to select an erosion resistant material (Table 1). Pumps running in fluids other than water and seawater are unlikely to be troubled by erosion. Pumps running at high speeds (above 200 ft/s tip speed) are very likely to suffer erosion damage if cavitation is present.

6. Cavitation in Water Turbines

6.1 DEFINITION AND PARAMETERS

Much of what has already been said also applies to water turbines,[32] but there are some essential differences. Firstly, water turbines are usually very large machines (approximately 10 metres diameter) and in consequence efficiency is very important. It is possible to provide sufficient submergence to prevent cavitation by putting the turbine at a low level. However, this is expensive in terms of excavation, and it is usual to allow some cavitation and accept the risk of erosion. Indeed, arrangements are often made for the regular welding up of erosion damage.

Secondly, a water turbine differs hydraulically from a pump in that the cascade of blades is an accelerating cascade (rather than a diffusing cascade as in a pump); minimum pressure thus occurs further back on the blade.

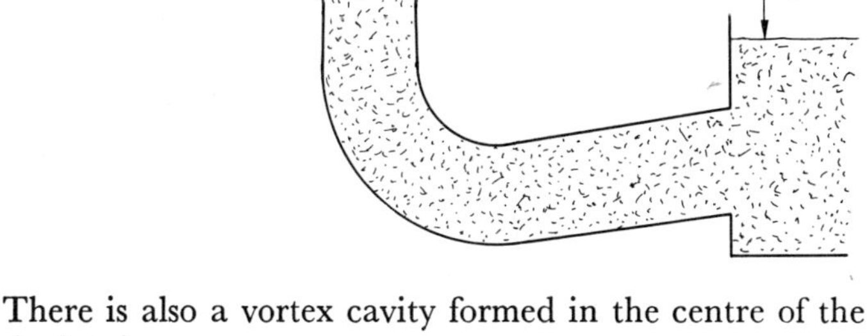

There is also a vortex cavity formed in the centre of the draft tube.

A typical overall turbine characteristic is shown in Fig. 27. From the cavitation point of view the most dangerous areas are at * where the turbine critical cavitation number is highest (full loads).

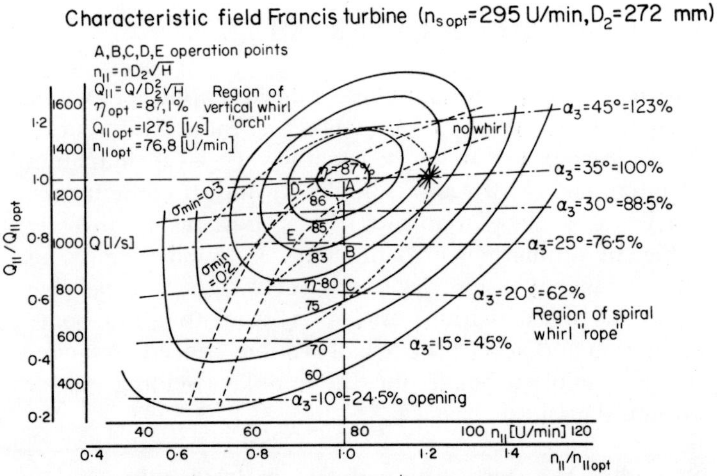

Fig. 27. Turbine characteristic showing points of cavitation.

The Thoma coefficient is commonly used for defining turbine cavitation coefficient

$$\sigma_{Th} = \frac{P_a - H_s - H_v}{H}$$

where H_s is height of centre line of turbine above tailrace, and H is net head across the turbine.

By analysis of test results on many turbines it is possible to make an empirical correlation of critical σ_{Th} against specific speed (Fig. 28), from which it is possible to derive general guidance for new design or new plant.

An approximate equation has been fitted to this by Karelin(33)

$$(\sigma_{Th})_{crit} = 0{\cdot}113 \text{ to } 0{\cdot}147\ \eta_T{}^{2/3}\ K_n{}^{4/3}$$

(or 1 to $1{\cdot}3 \times 10^{-4} N_{st}{}^{4/3}$).

Mosonyi(34) quoted for Francis turbines

$$(\sigma_{Th})_{crit} = 0{\cdot}12\ \eta_T K_n^2$$

(or $0{\cdot}032(N_{st}/100)^2$).

It is also possible to express these in terms of suction specific speeds. Karelin's figures convert to

$$K_{s_{crit}} = 5{\cdot}4 \text{ to } 4{\cdot}4$$

($s' = 385$ to 231 metric units), and Mosonyi to

$$K_{s_{crit}} = \frac{5{\cdot}2}{\sqrt{K_n}}$$

$\left(s' = \dfrac{27{\cdot}6}{\sqrt{N_s'}} \text{ metric units}\right)$.

Noskievic [35] shows a variable $K_{s_{crit}}$ from 4·5 to 5·8, the values shown in Fig. 28 (s' from 238 to 306).

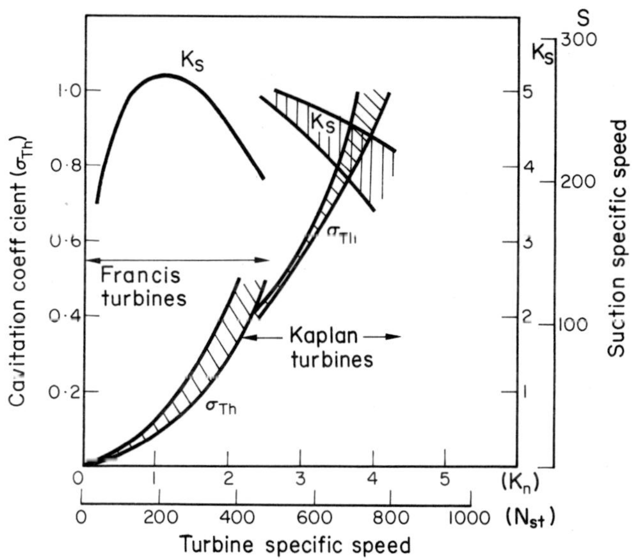

Fig. 28. Cavitation limits for turbines (Ref. 35).

Thus a conservative value of $K_{s_{crit}}$ for turbines would be about 4·4, and it should be remembered that this can be improved on by good design.

6.2 OPTIMUM BLADE DESIGN

As in the case of pumps, cavitation performance can be

expressed as a function of ζ_v a blade cavitation number, and ζ_c a loss coefficient in the discharge.

$$\sigma_{Th} = \zeta_v \frac{W_2^2}{2gH} + \zeta_c \frac{V_{2m}^2}{2gH}.$$

Noskievic gives $\zeta_c = 0{\cdot}6$ to $0{\cdot}75$ and ζ_v depends on the aerodynamic profile.

Alternatively, Nechleba[32] suggests

$$\text{critical } \sigma_{Th} = \frac{1}{2gH}[W^2\infty(1 + KC_L) - \eta_s C_4^2 - U^2 + 2UV_{U2} - (1 - \eta_s)(V_a^2 + V_{U2}^2)]$$

where $W\infty = (W_1 + W_2)/2$,

C_L is the lift coefficient (including cascade coefficient if necessary),

K is a cavitation factor greater than 1, and

η_s is draft tube efficiency.

Betz gave values of K and C_L

C_L	0·71	1·00	1·26	1·46
K	1·33	1·37	1·75	2·12

Another approach is to optimise the outlet velocity triangles with respect to σ_{Th}. This gives the result

$$\sigma_{Th_{opt}} = 0{\cdot}207 \frac{U_2^2}{2gH}$$

for $\zeta_c = 0{\cdot}3$ and $\zeta_v = 0{\cdot}2$ which gives some outlet swirl.

6.3 DRAFT TUBE

Vortex formation in the draft tube is a phenomenon present in turbines that has no parallel in pumps. When the turbine is operating at conditions away from the design or best efficiency point, the outlet flow is no longer axial and residual swirl is left in the draft tube flow. (These are in opposite directions depending on whether above or below the best efficiency point.) The effect of the residual swirl is to cause a large air-entraining cavity in the centre of the draft tube. Unfortunately this often oscillates or precesses around the draft tube, causing severe vibrations in the prototype. Fig. 27 shows the area of no whirl and vortex function on the turbine characteristic. The relation of the vortex shape to cavitation number for operation point B is shown in Fig. 29.[51]

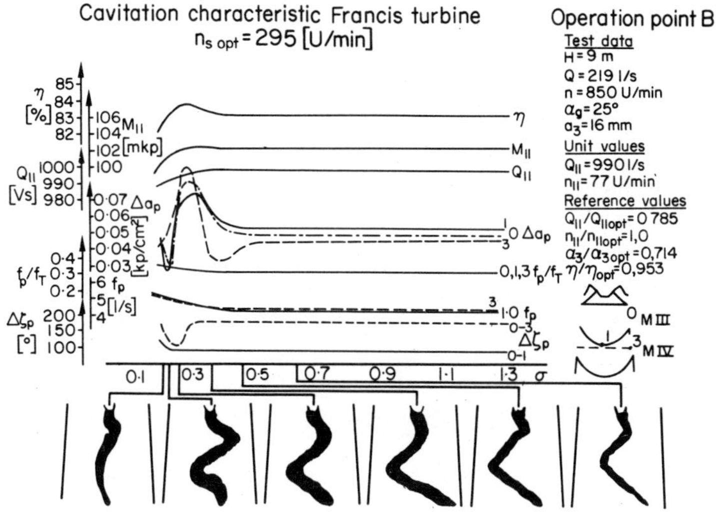

Fig. 29. Cavitation characteristics—Francis turbine—showing draft tube "rope".

Two methods are normally employed to alleviate the effects of these pulsating vortices. One is air injection into the low pressure core of the draft tube; this has a stabilising effect on the cavity vortex and stops the heavy vibration. The other is the placing of a cylindrical coaxial diffuser in the middle of the draft tube. This prevents the vortex precessing round the draft tube and stabilises it. It also has little effect on efficiency.

Such phenomena can be satisfactorily investigated in models, but the exact scaling laws are not yet known, so the results are mainly qualitative.

6.4 EROSION

Most large turbines are steel fabrications, although there are some steel castings. It is standard practice to weld on stainless steel overlays in the areas where erosion damage is likely to occur (as determined from model tests). It is also standard practice to repair such erosion damage at regular intervals (perhaps 1 or 2 years). Indeed, it is usual to include restrictions and penalties on erosion damage in contracts, stating that the erosion has to be less than a certain weight or volume loss in a given time. For instance, the amount of material loss from the runner during the first 10,000 hours shall not exceed 20, 60 or perhaps 100 lb of metal. In addition, it may be agreed that the metal thickness of the runner shall not be reduced due to pitting by more than $\frac{1}{8}$ inch maximum over any single area of 4 square inches or more. If the turbine fails to meet this guarantee the manufacturer shall repair the areas with 18/8 stainless steel welding and finish grinding to shape. These guarantees are made on the basis of previous experience and usually carry with them the manufacturer's penalty of making good the damage.

With regard to materials, the considerations of Section 3 apply equally well to turbines.

7. Propellers

7.1 EFFECTS AND PARAMETERS

Cavitation creates three main problems with respect to propellers; the effects on performance, erosion and vibration (including noise). Of these, the effect of cavitation on performance (thrust breakdown) has been most extensively studied. Design philosophy is to aim at either noncavitating or fully cavitating operation and to avoid partial cavity regions. In many cases, complete avoidance of cavitation is not possible and for this reason model tests are used to predict the performance in practice.

Four different types of cavitation are defined; bubble, streak, sheet and vortex cavities. *Bubble cavitation* is a group of separate cavities, each cavity being clearly defined. Erosion is often associated with this type. As the cavitation develops it becomes *streak cavitation*, consisting of long narrow sheets occurring often in parallel groups. *Sheet cavitation* forms as a single attached cavity with the upstream edge clearly defined. When it is longer than the blade it becomes supercavitation. Cavitation often forms also in the tip vortices, producing *vortex cavitation*, which is probably the most characteristic type of propeller cavitation (Fig. 30).

For a propeller, the cavitation number is

$$\sigma = \frac{P_1 - P_v}{\frac{1}{2}\rho W^2}$$

where P_1 is the pressure upstream of the propeller, and
W the relative velocity, at blade section.

A number of forms are used. P_1 can be defined either at the screw axis, or at point of minimum immersion; dynamic pressure may be related to speed of advance, speed of rotation at 0·7 radius or relative velocity as above.

Fig. 30. Propeller cavitation. Crown Copyright courtesy of the National Physical Laboratory.

Performance is usually quoted in terms of the

thrust coefficient $K_T = \dfrac{T_h}{\rho n^2 D^4}$

torque coefficient $K_Q = \dfrac{T_Q}{\rho n^2 D^5}$

advance coefficient $J = \dfrac{V_a}{nD}$

and efficiency $\eta = \dfrac{T_h V_a}{2\pi n T_q}$

where T_h = thrust, and
T_q = torque.

7.2 PERFORMANCE AND DESIGN

The pressure surface is often called the face and the suction surface the back of the propeller.

As already shown, cavitation causes a reduction in lift and increase of drag on a hydrofoil (Figs. 13–15). The propeller is affected in a similar manner and Fig. 31 shows what happens. As σ drops, the thrust, torque and efficiency all diminish, in manner similar to that typical of an axial pump. The effect is more marked at low values of the advance coefficient J (corresponding to increased incidence). The extent of cavitation is shown in the accompanying diagram, taken from tests on a model propeller in a water tunnel.

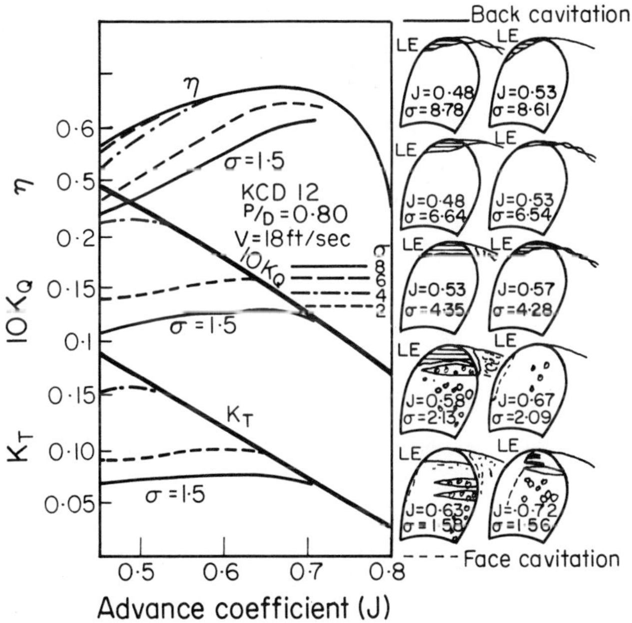

Fig. 31. Effect of cavitation on the performance of a propeller (Ref. 36).

In designing a propeller for a ship, a major problem is created by the fact that the propeller operates in a highly non-uniform flow due to the boundary layer and shaft angle. In a merchant ship the flow angle at 0·7 radius may vary from −8 to +2 degrees during each revolution. For a warship it is $-2\frac{1}{2}$ to $+1\frac{1}{2}$ degrees. Thus the propeller must have no cavitation over this operating range. A similar problem is caused by the fact that some propellers, such as those of tugs, trawlers, planing craft etc., have to operate over widely varying coefficients.

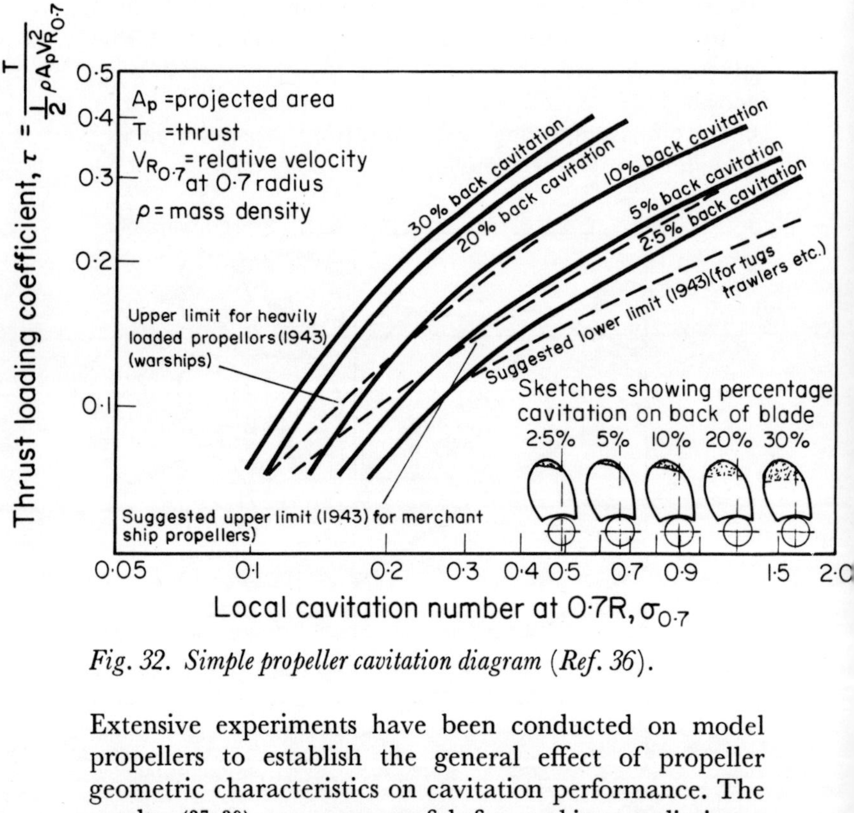

Fig. 32. Simple propeller cavitation diagram (Ref. 36).

Extensive experiments have been conducted on model propellers to establish the general effect of propeller geometric characteristics on cavitation performance. The results [37–39] are very useful for making preliminary calculations; they indicate upper limits on thrust loading coefficients and cavitation number (Fig. 32) where

$$\text{thrust loading coefficient} = \frac{\text{thrust}}{\frac{1}{2}\rho A_P W^2{}_{R0\cdot7}}$$

where A_P = projected blade area, and
$W_{R0\cdot7}$ = relative velocity at 0·7 radius.

$$\sigma = \frac{P_1 - P_V}{\frac{1}{2}\rho W^2{}_{0\cdot7}}$$

where P_1 is at propeller axis.

Some cavitation is allowable, about 5 per cent for merchant ship propellers and up to 10–20 per cent for warship propellers, depending on the loading. Above 20 per cent is not recommended as the performance is beginning to drop off in this region. From these data, estimates of cavitation inception and performance degradation of propellers can be made without detailed design calculations.

For higher speed craft where the limits have to be exceeded it is better to use a supercavitating propeller.

The design of subcavitating propellers is based on lifting line theory[40] (each blade is represented by a line vortex) to obtain the load distribution on the propeller, and lifting surface theory (each blade is represented by the distribution of vortices and sources) to give correction for the chordwise load distribution and for chordwise and spanwise thickness distributions. Allowance must be made for the variation of flow angle to the blades, due to the hull rake.

When the hydrodynamic design has been completed, the cavitation performance can be calculated by considering it as a two-dimensional section operating at the relative velocity with the appropriate incidence (which fluctuates). Such data are as shown in Figs. 13–15. It is also necessary at this point to consider strength, as thickness and chord length are both affected by strength, and they in turn have an adverse effect on cavitation. The design is then an iterative process between the often conflicting requirements of lift, strength and cavitation which is modified until a satisfactory compromise is reached.

7.3 SUPERCAVITATING PROPELLERS

For high speed craft, high propeller speeds are necessary to get the required thrust with a propeller of reasonable size. It is not possible to design for no cavitation, and a supercavitating propeller can be used, where a cavity is deliberately formed covering the back of the propeller and collapsing in the wake.

Fig. 33 shows the regions where ordinary propellers and supercavitating propellers should be used.

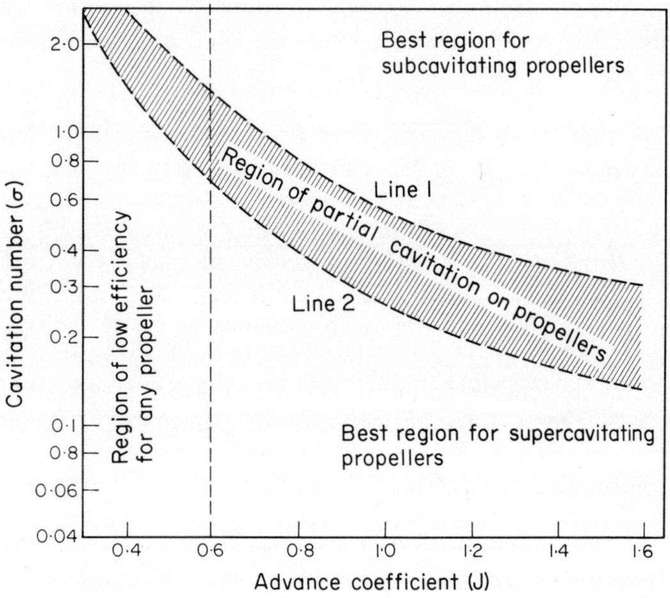

Fig. 33. Practicability of supercavitating propellers.

Supercavitating sections, as already shown, are wedge shaped. The thin leading edge is necessary and a thick trailing edge is useful for strength reasons. However, most craft have to operate at low speeds (entering and leaving harbour), so if possible the trailing edge should be made thin. The design problems are much more difficult than for ordinary propellers, because of the interdependence of cavity thickness on camber and incidence and of lift on cavity shape. Although several design methods have been suggested,[41, 42] none is as yet entirely adequate.[43, 44]

7.4 EROSION

Chemical and cathodic erosion are hazards which must be considered in addition to cavitation erosion in the marine field. The designer meets these hazards by choice of suitable material. Cathodic protection is often used on ships and there is some evidence to show that this reduces erosion damage slightly.[45] As indicated in Section 3, the combination of erosion and corrosion accelerates loss of metal as the erosion removes the oxide films that often develop in a corrosion process (Fig. 34). For this reason marine propellers are usually made from materials with good corrosion and erosion resistance, such as aluminium bronze or manganese bronze.

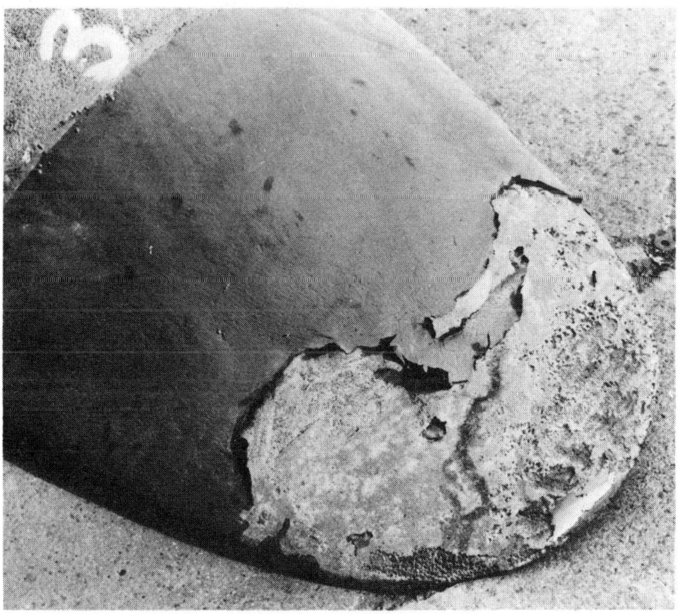

Fig. 34. Cavitation erosion damage on a propeller. Crown Copyright courtesy of the National Physical Laboratory.

As indicated previously, it is still not possible to predict whether erosion will accompany cavitation. In propeller design, therefore, everything possible is done, especially by using model tests, to avoid cavitation in order to eliminate the highly-damaging root or tip cavities. Erosion-resistant materials are used in case the predictions are not exact. (The prototype propeller may not always be geometrically identical to the model.) It must be accepted that propeller repair may be necessary if excessive damage occurs; this applies particularly to warships and high speed craft.

7.5 OTHER MARINE DEVICES

Many other marine devices may suffer cavitation problems; they include hydrofoils, struts, torpedoes, propeller nacelles, sonar domes, stabilisers. Most of these can be designed with the aid of theory or model experiments, to minimise cavitation (Section 4). Many suffer erosion damage, which can be partially alleviated by material selection as already described (Section 3). In this particular field a great deal of attention has been paid to the use of protective overlays of soft (rubber, plastics) or hard (stainless steel) materials. In general, these overlays have not been successful due to the difficulty of getting satisfactory adhesion to the parent surface.[46–48]

8. Cavitation Testing

It would be very difficult to carry out cavitation tests on the prototype of a machine, so scale models are usually used for such tests. The main requirement is to measure the machine's performance (whether pump, turbine, propeller, torpedo or hydrofoil) and at the same time to vary pressure to cover the range from no cavitation to performance breakdown. It is usual to use a closed-loop water tunnel (Fig. 35); in addition to having facilities for measurement, this commonly has transparent windows for viewing cavitation. The essential elements of such a tunnel are a working section where the object under test is mounted, dynamometers or pressure measurements to analyse its performance, a variable speed pump to circulate the flow, and a pressure control.

The tunnel shown in Fig. 35[(52)] is normally used for water turbine testing. The model turbine (500 mm) is mounted at the top and the water flows from a contraction and flow straightener to ensure a uniform velocity profile and low turbulence. The flow from the draft tube goes into a settling tank and then passes down through a flowmeter to the pumps 10 m below (to avoid cavitation in the pumps). The pumps are situated on a reabsorber tank, which is used to reabsorb gas bubbles released during cavitation testing. The water passes up and down four times inside the reabsorber whilst under pressure and the gas (mostly air) is forced back into solution. After leaving the pumps the water recirculates up to the turbine.

The overall circuit has a pressure control to vary the absolute pressure at the turbine from near zero to 4 atmospheres. There is a refrigerator system to cool the circuit and a deaerator to vary the air content of the water. The circuit can be changed to an open circuit, so the flowmeter can be calibrated by a weigh tank. The turbine drives an electric dynamometer, so torque and speed can be measured as well as head and flow to give efficiency.

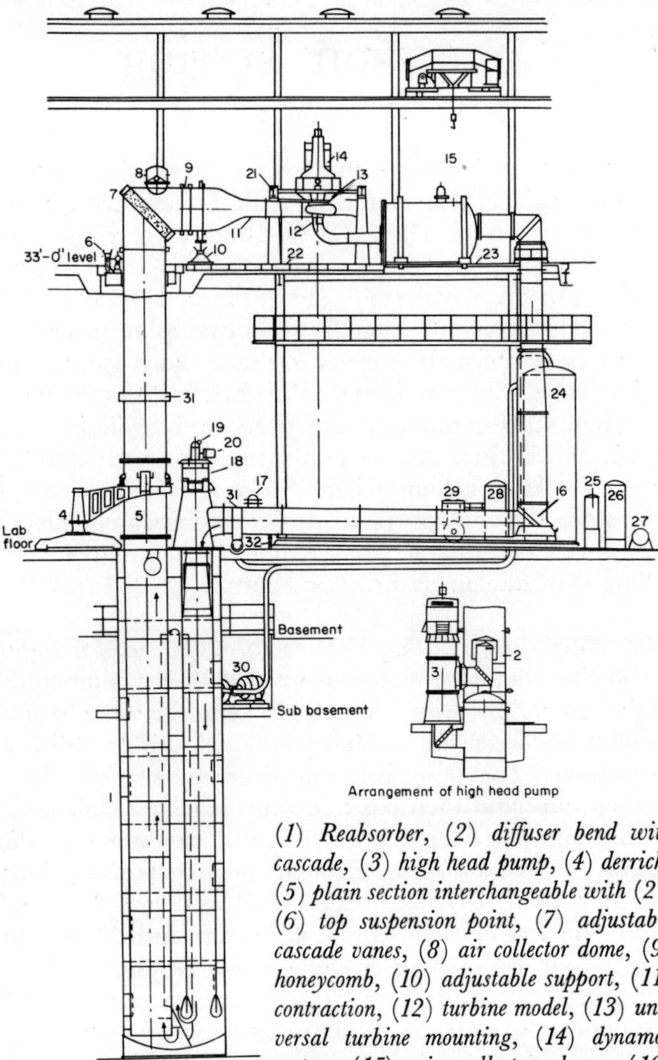

(1) Reabsorber, (2) diffuser bend with cascade, (3) high head pump, (4) derrick, (5) plain section interchangeable with (2), (6) top suspension point, (7) adjustable cascade vanes, (8) air collector dome, (9) honeycomb, (10) adjustable support, (11) contraction, (12) turbine model, (13) universal turbine mounting, (14) dynamometer, (15) air collector dome, (16) cascaded bend, (17) air collector dome, (18) variable pitch propeller pump, (19) pitch changing mechanism, (20) speed control, (21) dynamometer bridge, (22) removable bedplate, (23) removable hatch, (24) deaerator vessel, (25) pressure control vessel, (26) vacuum, intermediate and high pressure vessels, (27) heat exchanger, (28) vacuum vessel, (29) vacuum pump, (30) circulating pump, (31) deaerator connection, (32) tunnel drain.

Fig. 35. Variable-pressure water turbine test rig, closed circuit.

For cavitation tests the turbine would be run at constant flow and power and the pressure would be dropped until cavitation occurred and eventually the performance dropped off.

Such a water tunnel can be used for the testing of hydrofoils, torpedoes or other bodies by replacing the turbine and settling tank by a working section. The working section would have transparent windows for viewing purposes and possibly force-measuring gear and pressure-measuring equipment for the hydrofoil. The section may have slotted walls to eliminate boundary layer interference.

A similar rig is used for propeller testing, the propeller being mounted in the working section and driven usually by a shaft from the upstream side. The propeller shaft would have both torque and thrust measuring dynamometers. Sometimes the propeller drive is arranged so it

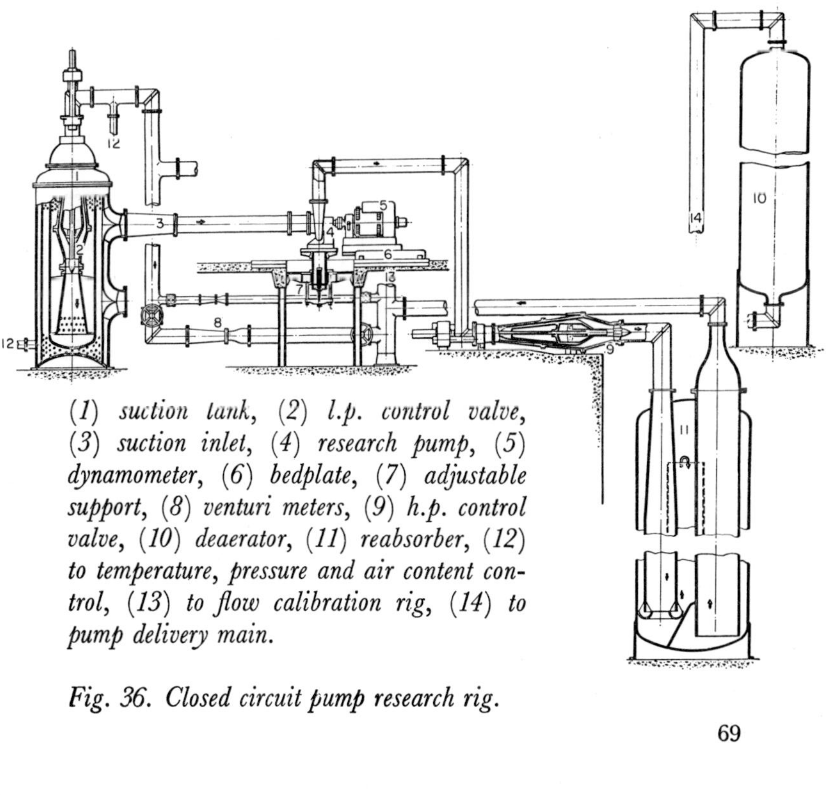

(1) suction tank, (2) l.p. control valve, (3) suction inlet, (4) research pump, (5) dynamometer, (6) bedplate, (7) adjustable support, (8) venturi meters, (9) h.p. control valve, (10) deaerator, (11) reabsorber, (12) to temperature, pressure and air content control, (13) to flow calibration rig, (14) to pump delivery main.

Fig. 36. Closed circuit pump research rig.

can be inclined to be more representative of ship applications. The flowmeter would not be required in such an application, the velocities in the working section being measured by pitot tube.

For testing pumps, which provide their own driving power, a simpler tunnel can be used (Fig. 36). Alternatively, an open circuit test could be used with the pump sucking water from an open tank, in which the level can be dropped to induce cavitation in the pump.

Methods of testing materials for erosion resistance have been described in Section 4.

9. Ultrasonic and Biological Cavitation

Cavitation can be produced by sound waves. Those of the highest ultrasonic frequencies, from 10 kHz to 10 MHz, are usually used in practice, as they create the smaller bubble sizes. The waves produce an alternating pressure field in the liquid and the bubbles grow and collapse in a cyclic manner. Ultrasonic techniques are commonly used to investigate the physics of cavitation as they provide effective control of bubble size and distribution. The effect on the inception of cavitation of such variables as temperature, viscosity, pressure and gas content has been investigated by these means.

9.1 CHEMICAL PROCESSES

Ultrasonically-induced cavitation can be used to assist manufacturing processes by performing such tasks as cleaning and promoting chemical reactions, degassing liquids, and emulsification. In general, the value of ultrasonic cavitation depends on one or two effects. The resonant bubbles can act as a mixer, increasing the area of contact between two liquids or a liquid and boundary surface. Cleaning processes can be assisted in this way, and immiscible liquids may be emulsified by the vigorous disturbance.

A second effect of ultrasonically-induced cavitation is the initiation of reactions that would not otherwise occur, more especially those which take place in aqueous media. Webster(49) cites a number of chemical processes that are initiated or accelerated by ultrasonic cavitation. For instance, solutions of polymers may decrease in viscosity when subjected to ultrasonic waves of high intensity, due to the rupture of the chemical bonds of the polymer chain.

9.2 BIOLOGICAL AND MEDICAL ASPECTS

Cavitation can have a destructive action on red blood cells, yeast cells and bacteria, and thus can be used in the extraction of cell contents. Enzymes have been extracted by these means, those of lower molecule weight being most suitable.

Removal of viruses from infected tissue has also been achieved. At low intensities growth is stimulated, but as the intensity increases, destruction sets in at a certain threshold and increases thereafter. The death rate increases with both time and temperature. The mechanism of bacterial destruction is thought to be due either to the effect of cavities inside the bacteria, or alternatively to the production of hydrogen peroxide in the water. The removal of dissolved gas and the alternating pressure certainly play some part in virus destruction.

Cavitation in blood has been blamed for physical damage to human organs. Blood flows in arteries and veins differ from most liquid flows, because they are markedly pulsating and are contained in an elastic pipe system. It has been suspected that damage to the heart valves may be caused by cavitation. Similarly, cavitation at the junction of arteries may cause thickening of the walls and result in arteriosclerosis.

Cavitation has been blamed for killing off fish. In many hydroelectric schemes fish are often carried down with the water to the turbines. Most of the fish safely negotiate the passages between the turbine blades, but in some stations many fish are killed. The fact that they are not mutilated indicates that they are not hit by the turbine blades; they succumb to the low pressures and steep pressure gradients caused by cavitation on the blades.

10. Future Trends

The greatest immediate research needs in the area of cavitation are to be able to predict if a machine or component is cavitating under certain fluid conditions, and to predict whether or not erosion will take place when a particular material is used. If erosion takes place, it is necessary to know the rate at which it occurs. These problems may not be solved for a decade or so, but their eventual solution will allow us to live with cavitation instead of avoiding it or fighting it.

It is probable that increasing use of supercavitating pumps will be made permitting the use of higher pump speeds and smaller pumps, with lower installation and maintenance costs. Initially these high speed pumps will be in pumps of the smaller sizes, but there will be a gradual increase of size as time goes on.

Water turbines of improved cavitation characteristics will be developed gradually. Supercavitating turbines are feasible (models have been built), but their moderate efficiency makes them unlikely contenders in the large schemes. However, the reduced size and cost of small turbines may make them useful for river or tidal applications.

In the case of ships' propellers, the avoidance of cavitation will continue to be a high priority consideration. However, high speed craft will need supercavitating propellers which must also be capable of operating at low speeds. Techniques and materials capable of minimising the effects of cavitation will therefore be important. Protective coatings especially have particular appeal in the marine field, provided satisfactory adhesion techniques can be developed. This applies equally to hydrofoils and supercavitating hydrofoils, whether used as lifting foils in hydrofoil boats, or as stabilisers or hydroplanes. For ship propulsion, development may be in the direction of ducted propellers or pump jet systems; these may change the type of cavitation problems, but they will not eliminate them.

It is likely that more resistant materials will be developed combining hardness and ductility, such as chrome manganese stainless steels. Plastics too show interesting prospects, although none so far has proved entirely satisfactory.

The future will see progress too in the useful applications of cavitation. Its uses in cleaning and in emulsification will no doubt be extended, and it could become the basis of surface finishing or machining processes. In this connection, control of cavitation remains the major problem. Hydraulically, cavitation occurs in a random and intermittent fashion. Ultrasonically, much better control can be exercised, but it is difficult to inject reasonable amounts of power through generators that are basically very inefficient.

The ability of liquids to withstand high tensions could be an extremely useful characteristic. Unfortunately, the high purity and the need for pre-pressurisation have prevented this property from being put to good use outside the laboratory. A potential application, for example, might be in a "packaged hydraulic circuit" as used on a rocket engine, where the liquid could be pretreated before being put into the circuit. The expense of cleaning and filtering, however, would probably outweigh the advantages of such a system.

REFERENCES

1. HARVEY, E. N., MCELROY, W. D. and WHITELEY, A. H. "On cavity formation in water." *J. appl. Phys.*, **18,** 162–72, 1947.
2. HAYWARD, A. T. J. "New law for liquids, don't snap stretch." *New Scient.*, 196–9, 29 January 1970.
3. PLESSET, M. S. "The tensile strength of liquids", Cavitation state of knowledge. *ASME*, 15–25, 1969.
4. WHEELER, W. H. "Indentation of metals by cavitation." *ASME Trans.*, Series D, **82,** N1, 184–94, 1960.
5. RIPKEN, J. F. and KILLEN, J. M. "Gas bubbles: their occurrence, measurement and influence in cavitating testing." *Proc. IAHR Symp.* Japan: Sendai, 1962.
6. PEARSALL, I. S. and MCNULTY, P. J. "Cavitation inception and efficiency of a two-dimensional diffuser." *MERL Fluids Report* **36.** Glasgow: NEL, 1956.
7. MANSFIELD, J. N. "Effect of head and air content on the cavitation performance of a turbine." *ASME Cavitation Forum,* 35, 1967.
8. SCHOENEBERGER, W. "Cavitation tests in radial pump impellers." Docter thesis TU Damstadt, 1965—see reference 50.
9. SCHMUGLYAKOV, L. S. "Relation between cavitation coefficient and air content for a turbine." DSIR Translation 441. *Energomashinostroenic* **11–14,** N5, 1956.
10. PARKIN, D. R. and HOLL, J. W. "Incipient Cavitation Scaling Expts for hemispherical and 1·5 calibre Ogive nosed bodies." Pennsylvania State Univ. ORL Report, 7958-264, 1954.
11. HOLL, J. W. and WISLICENUS, G. F. "Scale effects on cavitation." *J. bas. Engng, Trans. ASME,* Series D, **83,** 385–98, 1961.
12. KERMEEN, R. W., MCGRAW, J. T. and PARKIN, B. R. "Mechanism of cavitation inception and related scale effects problem." *Trans. ASME,* **77,** 533–41, 1955.
13. CALEHUFF, G. L. and WISLICENUS, G. F. "ORL investigation of scale effects on hydrofoil cavitation." Pennsylvania State Univ. ORL Report TM 19-4212-03, February 1956.
14. PEARSALL, I. S. "Acoustic detection of cavitation." Symp. Vibrations in hydraulic pumps and turbines. *I.Mech.E. Proc.,* **181,** Pt 3A, 1966–7.

15. PEARSALL, I. S. and MCNULTY, P. J. "Comparison of cavitation noise with erosion." *ASME Cavitation Forum*, 6–7, 1968.
16. HOBBS, J. M., LAIRD, A. and BRUNTON, W. C. "Laboratory evaluation of the vibratory cavitation erosion test." NEL Report 271. National Engineering Laboratory, 1967.
17. BRUNTON, W. C., HOBBS, J. M. and LAIRD, A. "Investigation of a cavitating film erosion test." NEL Report 431. National Engineering Laboratory, 1969.
18. HOBBS, J. M. and LAIRD, A. "Pressure, temperature and gas contents effects in the vibratory cavitation test." NEL Report 438. National Engineering Laboratory, 1969.
19. HERRIG, L. J., EMERY, J. C. and ERWIN, J. R. "Systematic two-dimensional cascade tests of NACA 65 Series compressor blades." NACA Tech. Note 3916, February 1957.
20. KERMEEN, R. W. "Water tunnel tests of NACA 4412 and Walchner profile 7 hydrofoils in non-cavitating and cavitating flows." Cal. Tech. Hyd. Lab. Report 47–5, 1956.
21. NUMACHI, F. and MURAI, H. "Cavitation tests on hydrofoil profiles arranged in cascade (1st Report)." *Report 13. Inst. High Speed Mech.* Tohuku Univ., vol. 2, N13, 1–44.
22. WADE, R. B. and ACOSTA, A. J. "Investigation of cavitating cascades." *J. bas. Engng*, Series D, **89,** 693–706, December 1967.
23. PARKIN, B. R. "Experiments on circular arc and flat plate hydrofoils." *J. Ship Res.*, **1,** 34–56, 1958.
24. DULLER, G. A. "On the linear theory of cascades of supercavitating hydrofoils." NEL Report 218. National Engineering Laboratory, 1966.
25. WOODS, L. C. and BUXTON, G. H. L. "The theory of a cascade of cavitating hydrofoils." *Quart. J. Mech. and Appl. Maths*, 387–402, 1966.
26. Standards of the Hydraulic Institute. Hydraulic Institute, New York, 11th Ed., 1965.
27. PEARSALL, I. S. and SCOBIE, G. "Pumps for low suction pressures." Paper No. 2.—Symposium on Pumping Problems. I.Chem.E. and I.Mech.E. Swansea, 25 March, 1970.
28. STEPANOFF, A. J. "Pumps and Blowers." John Wiley & Sons, 1965.
29. WARD, T. and SUTTON, M. "A review of the literature on cavitation in centrifugal pumps with various liquids"—TN 892. BHRA, Cranfield, April 1967.

30. SUTTON, M. "The inducer: A means of improving the suction performance of centrifugal pumps." BHRA RR889. BHRA, Cranfield, 1967.
31. PEARSALL, I. S. "Improved NPSH performance by the use of cavitating pumps." BHRA Conference, September 1967. Paper SP930.
32. NECHLEBA, M. "Hydraulic turbines." Constable & Co, 1957.
33. KARELIN, V. J. "Kavitacionnye javlenija v centrobeznyel." Masgiz Moshva, 1963.
34. MOSONYI, E. "Water power development." Hungarian Academy of Sciences, Budapest, 1960.
35. NOSKIEWICZ, J. "Kavitace". Academia, Prague, 1969.
36. BURRILL, L. C. and EMERSON, A. "Propeller cavitation, further tests on 16 inch propeller models in the Kings College cavitation tunnel." *Nth East Coast Inst. Engng and Ships*, **79,** 295–320, 1962.
37. LINDGREN, H. "Model tests with a family of 3 and 5 bladed propellers." Statens Skeppsprovingsansalt, Goteberg Report 47, 1961.
38. TANIGUCHI, K., TANIBAYASHI, H. and CHIBA. "Investigation into the propeller cavitation in oblique flow." Mitsubishi Expt Tank Reports 1800 and 2221, 1964 and 1966.
39. EMERSON, A. and SINCLAIR, L. "Propeller cavitation: Systematic series tests on 5 and 6 bladed propellers." *Trans. Soc. Nav. Arch. and Marine Engng,* **75,** 224–67, 1967.
40. MORGAN, W. B., SILOVIC, V. and DENNY, S. B. "Propeller lifting surface corrections." *Trans. Soc. Nav. Arch. and Marine Engng*, **76,** 1968.
41. TACHMINDJI, A. J. and MORGAN, W. B. "The design and estimated performance of a series of supercavitating propellers." Proc. 2nd ONR Symp. on Naval Hyd. ACR38, 488–532, 1958.
42. HECKER, R., PECK, J. G. and MACDONALD, N. A. "Experimental performance of TMB supercavitating propellers." DTMB Report 1432, January 1969.
43. COX, G. G. "Supercavitating propeller theory. The derivation of induced velocity." 7th ONR Symp. on Naval hydrodynamics. Rome, August 1968.
44. ENGLISH, J. "An approach to the design of fully cavitating propellers." ASME Symposium cavitation in Fluid Machinery, *ASME,* 7 November, 1965.
45. PLESSET, M. S. "On cathodic protection in cavitation damage." *Trans. ASME J. bas. Engng,* 808–20, 1960.

46. LICHTMAN, J. Z. "Cavitation erosion performance and related properties of cured sheet elastomeric coating systems." *J. of Materials*, **2,** N3, 638–60, 1967.
47. LICHTMAN, J. Z. and KALLAS, D. H. "Erosion resistance of coatings." *Materials protection*, **6,** N4, 40–45, April 1967.
48. NAKAJO, T., KUBO, H. and HAMADA, T. "On the cavitation erosion resistance of coatings." *J. Japan Soc. of Mat. Sciences*, **14,** N136, 54–60.
49. WEBSTER, E. "Cavitation", *Ultrasonics*, **1,** p. 39–48, January/March 1963.
50. ROBERTSON, J. M. and WISLICENUS, G. F. "Cavitation state of knowledge." *ASME*, 1969.
51. MOLLENKOPF, G. and RAABE, J. "Measurements of fluctuations of velocity and pressure in the draft tube of a Francis turbine." IAHR Symposium, Part 1. Stockholm, August 1970.
52. WINTERNITZ, F. A. L. and MANSFIELD, N. "Variable pressure water tunnel test rig at NEL." NEL Report No. 92. Glasgow, East Kilbride: National Engineering Laboratory, 1963.
53. JARMAN, P. "Measurements of sonoluminescence from pure liquids and some aqueous solutions." *Proc. Phys. Soc.*, **73,** 628, 1959.
54. HARRISON, M. "Experimental study of single bubble cavitation noise." *J. Acoust. Soc. Amer.*, **24,** P776, 1952.
55. THOMA, D. "Die Kavitation bei Wasser-Turbinen." VDI Verlag, Berlin, 65–74, 1926.
56. REYNOLDS, O. "The causes of the racing of the engines of screw steamers. Investigated theoretically and by experiment." *Tr. Inst. Naval Arch. V*14 *Sc. Papers*, **1,** p. 56-7, 1873.
57. RICHARDSON, A. "The evolution of the Parsons Steam Turbine." *Engineering*, 1911.

INDEX